Retail Best Practices and Supervisor's Guide to Food Safety and Sanitation

David McSwane, H.S.D.
Richard Linton, Ph.D.
Nancy R. Rue, Ph.D.

Production services and Development by
Learnovation, LLC

PEARSON
Prentice Hall

Upper Saddle River, NJ 07458

Editor-in-Chief: Steve Helba
Aquisitions Editor: Vernon Anthony
Managing Editor: Mary Carnis
Production Editor: Brian Hyland
Director of Manufacturing and Production: Bruce Johnson
Design Director: Cheryl Asherman
Senior Design Coordinator: Miguel Ortiz
Editorial Assistant: Ann Brunner
Desktop Publisher: Karen J. Hall
Developmental Editor: Anna Graf Williams
Assistant Development Editor: Dave Morrow
Cover Illustration: John Wise

Pearson Education LTD.
Pearson Education Singapore, Pte. Ltd.
Pearson Education Canada, Ltd.
Pearson Education–Japan
Pearson Education Australia PTY, Limited
Pearson Education North Asia Ltd.
Pearson Educación de Mexico, S.A. de C.V.
Pearson Education Malaysia, Pte. Ltd.

PEARSON Prentice Hall

10 9 8 7 6 5 4 3 2

ISBN 0-13-064842-6

Note from FMI

Food safety has always been a paramount priority for the food industry. Today it is even more so due to improved methods of detecting safety hazards and to increased consumer concerns about both food safety and security. As the last link in the global supply chain that delivers food to consumers, retailers play a special role in protecting public health.

In recent years, the industry, government and scientific community have designed effective programs to safeguard the food supply. Some are required by law, and others are embodied in good retail practices, standard operating procedures and training that teaches lasting lessons. The programs may vary, but all feature the same basic principles, including sanitation, employee health and hygiene - all aimed to prevent contamination before customers purchase food products.

The guidelines in the following pages are one of several educational tools developed through the cooperative efforts of the Food Marketing Institute and Prentice Hall. The authors of this manual are among the world's leading food safety experts and educators. They bring decades of experience training retail food handlers and managers how to systematically prevent contamination.

This training manual identifies best practices that prevent foodborne illnesses at the retail level. Knowing these is only the beginning in fully understanding food safety. Apply this knowledge daily each time you prepare, handle, display or offer food. These practices will help ensure that your customers become loyal patrons, confident your food is safe and wholesome.

Tim Hammonds

President and CEO

Food Marketing Institute

Contents

Chapter 5 Facilities, Equipment, and Utensils 136

PREFACE

Supervisors and managers of retail food establishments take on a great burden of keeping food and food products safe from the moment it enters the establishment. There are many opportunities for food to be contaminated between production and consumption. Food can be contaminated at the farm, ranch, orchard, or in the sea. Food also can be contaminated at food processing plants and during transport to retail food establishments. Finally, food can be contaminated during the last stages of production, at retail establishments, and by patrons in their homes. The importance of keeping food safe cannot be underrated. Retail food supervisors and managers can use the *Retail Best Practices and Supervisors Guide to Food Safety and Sanitation* to assist them in this daunting task.

Food safety is especially critical in retail food establishments because this may be the last opportunity to control or eliminate the hazards that might contaminate food and cause foodborne illnesses. Even when purchased from inspected and approved sources, ingredients may contain contaminants when they arrive at a retail food establishment. It is important to know how to handle ingredients safely and how to prepare food in a manner that reduces the risk of contaminated food being sold to your patrons.

The Food and Drug Administration (FDA) is one of the federal agencies responsible for protecting our nation's food supply. The Agency recognizes the importance of food safety in retail food establishments. In fact, the agency's 2001 Food Code recommends that retail food managers be able to demonstrate knowledge in food safety. This knowledge will be very helpful to managers as they create and implement food safety management programs within their operations.

Food safety in retail food establishments begins with managers who are knowledgeable about food hazards and who are committed to implementing proper food-handling practices in their facility. It also requires properly trained food workers who understand the essentials of food safety and sanitation and who will not take short cuts when it comes to food safety.

The authors of this textbook have been training retail food managers and employees for over 25 years. Many excellent resources are available for this type of training. However, the authors wanted to create a book that was customized for the retail food industry and would meet the training needs of supermarkets, superstores, food warehouses, limited assortment stores, convenience stores, food warehouses, and other types of traditional and nontraditional stores.

The Best Practices materials have been proven effective for teaching food safety and sanitation to many different audiences. The authors recommend the textbook and supplemental materials in-class and home study courses to prepare retail food establishment managers to take a national food protection manager certification examination.

One of the most important tasks you face is to train and supervise food workers. Your knowledge is useless if you do not teach employees the correct way to handle food. Learn to recognize any break in standard operating procedures that might endanger food safety. Avoid the problems related to embarrassment, loss of reputation, and financial harm that accompany a foodborne disease outbreak.

Some state and local jurisdictions have passed legislation to require certification of one or more managers in each retail food establishment. Other jurisdictions are considering doing the same. Most certification programs require candidates to pass a written examination to demonstrate knowledge of food safety and sanitation principles and practices. Some jurisdictions require completion of a food safety course before taking the exam.

There is growing support for a nationally recognized examination and credential for retail food protection managers. The Conference for Food Protection (CFP) recognizes food protection manager examinations from several providers. Contact the CFP at www.foodprotect.org for more information about the test recognition process and the various providers who have one or more forms of their examination recognized by the Conference.

The team of McSwane, Linton, and Rue wanted to produce a book to accompany the FDA's 2001 Food Code. We trust that you will find this text accurate, comprehensive, and, most of all, useful.

David Z. McSwane, H.S.D., REHS, CFSP, is an Associate Professor of Public and Environmental Affairs at Indiana University. He has over 25 years of experience in food safety and sanitation working in state and local regulatory agencies and as a consultant to the food industry. Dr. McSwane has published numerous articles and presented papers on a variety of subjects related to food safety.

Dr. McSwane is a nationally recognized trainer in food safety and sanitation. He has taught courses at the university level and for regulatory agencies, retail food establishments, food industry trade associations, vocational schools, and environmental health associations throughout the United States. Dr. McSwane is a recipient of the Walter S. Mangold award. This is the highest honor bestowed by the National Environmental Health Association. He has been a correspondent for the Food Protection Report and is a member of the Environmental and Public Health Council at Underwriters Laboratories Inc.

Richard Linton, Ph.D., is Professor of Food Safety at Purdue University. His expertise is in the development and implementation of food safety and food quality programs, specifically in Hazard Analysis Critical Control Point (HACCP) systems. Dr. Linton spent twelve years working in retail food establishments. In recent years, he has educated food workers and managers from all segments of the industry throughout the nation and the world.

Dr. Linton has developed several different types of food safety training programs for retail food managers and workers. He also works closely with the retail food industry on research projects that help improve the quality and safety of the food they serve.

Nancy Roberts Rue, Ph.D., R.N., has a background in teaching in technical education. Her doctorate is in educational leadership and curriculum and instruction from the University of Florida. With thirty years of experience in higher education and evaluation at Indiana University and St. Petersburg Junior College, she is dedicated to the task of building educational materials that meet the needs of those who want to learn. She is now an independent writer and consultant on training and development techniques.

Learnovation®, LLC has provided expert instructional design and developmental editing for the *Retail Best Practices and Guide to Food Safety and Sanitation.* Recognized for their team of experts with years of experience in instructional design, they assisted in the customization of this book and developed tables, charts, key point boxes and illustrations that allow for ease of learning.

The Retail Best Practices and Supervisors Guide to Food Safety and Sanitation is a part of a series of books and training materials that apply directly to retail food establishments such as supermarkets and convenience stores. This is a quick and easy read, full color guide including the "must know" information about food safety and sanitation in retail food operations.

In addition to *The Retail Best Practices and Supervisors Guide to Food Safety and Sanitation*, the food safety series, endorsed by the Food Marketing Institute includes:

- ▲ *The Retail Best Practices and Guide to Food Safety and Sanitation*: a comprehensive overview of all food safety principles that impact retail food establishments
- ▲ *The Retail Best Practices and Quick Reference Guide to Food Safety and Sanitation*: an easy to read illustration rich text designed for the line worker. This concise reference focuses on the key areas of personal hygiene, time and temperature abuse, cross contamination as well as sanitizing and cleaning practices.
- ▲ *The Retail Best Practices and Quick Reference Guide to Food Safety and Sanitation:* an easy to follow, bulleted training and teaching guide with a PowerPoint slide presentation full of effective illustrations and photos that

help drive food safety points home. The kit also includes an image bank of photos and illustrations for customizable presentations and printable full color posters that address important food safety principles.

The authors wish all readers success in their food safety and sanitation activities. Regardless of where you work, you must always remember--foodborne illness is a preventable disease. Follow the basic rules of food safety and you will enjoy a satisfying career in the retail food industry.

Control Point Icons

Throughout the Retail Best Practices and Supervisors Guide to Food Safety and Sanitation you will encounter control point icons in the margins of the text. Control point icons are placed next to text indicating moments where action might be taken by a supervisor or an employee to correct, or prevent a possible food safety issue. Control point icons can be found throughout the text for each of the following important food safety areas:

▲ **Proper Cleaning and Sanitizing**

▲ **Preventing Cross Contamination**

▲ **Avoiding Time and Temperature Abuse**

▲ **Proper Personal Hygiene**

ACKNOWLEDGMENTS

The authors wish to thank their families and colleagues who, through their support and patience, helped us with the first edition of this book. John Wise is a tremendous asset to this project. His drawings enable us to present information that is technical and sometimes tedious in a way that make it easy to understand. Karen J. Hall, David Morrow, and George Williams from Learnovation, desktop publishing and developmental editors, gave us the technical support needed to make the book come alive.

Special thanks goes to Tom Compbell and Vincent P. Walter, the project photographers for their innovative and inspired services. Scott Alkinburgh of Marsh Supermarkets and Don Loutzenhiser of Martin's Supermarkets each graciously shared their knowledge and facilities with us during the writing of the book. A heartfelt acknowledgement goes to Vernon Anthony, the acquisitions editor at Prentice Hall, for his vision and support during this process.

A grateful thanks to goes to Timothy Weigner and Laurie Williams, Food Safety Programs at Food Marketing Institute, for their technical expertise and insight into supermarket food safety programs.

We would like to acknowledge and recognize the supermarket leaders who provided expertise and review of the program as volunteers. They include:

Ruth Yong
The Kroger Co.

Cory Hedman
Hannaford Bros. Co.

Robert Brown
Shaw's Supermarkets, Inc.

Dave Pettigrew
Spartan Stores, Inc.

Monica Martin
Farm Fresh, Inc.

Garey Walker
Giant Food, Inc.

Scott Alkinburg
Marsh Supermarkets

Krag Swatz
Lund Food Holding, Inc.

Michael Hillyer
Wal Mart Stores, Inc.

Don Loutzenhiser
Martin's Super Markets

George Sherback
The Stop & Shop Supermarket

Gillian Kelleher
Wegmans Food Markets, Inc.

Michael Magner
Price Chopper Supermarkets

Cas Tryba
Big Y Foods, Inc.

Learn How To:

- Recognize how food safety and sanitation practices prevent foodborne illness in retail food establishments.
- State the problems caused by foodborne illness for individuals who become ill and the retail food establishment blamed for the incident.
- Identify trends in menus and consumer use of food products prepared in retail food establishments.
- Describe the role of government (federal, state, and local) in retail food safety.
- List the types of retail food establishments identified in this text and the influence of the *FDA Food Code* and state or jurisdictional codes on these operations.
- Define the term Hazard Analysis Critical Control Point (HACCP) as applied in food safety management.
- Recognize the need for food protection manager certification.

CHAPTER 1

Retail Food Safety

Multi-state food recall over contaminated beef...

The top story on the 6 o'clock news last night involved a multi-state food recall. Over 24 million pounds of ground beef was being recalled from a major supermarket chain because it was contaminated with bacteria called E. coli *O157:H7. At the time of the recall, 313 people had reported flu-like symptoms and some indicated they had experienced extreme bloody diarrhea. Two children died. All the contaminated beef was processed at XYZ meat processors. The contaminated lots of beef included frozen beef patties, frozen beef burritos, and fresh ground beef under the XYZ label containing the lot codes: 78536, 78439, 78339, and 78206. What impact could this recall and associated foodborne illness and deaths have?*

New Challenges Present New Opportunities

The food industry is one of America's largest enterprises. It employs about one-quarter of the nation's work force and produces 20% of America's Gross Domestic Product (GDP). Billions of dollars worth of food are sold each year. Americans have made food a prominent part of their business and recreational activities.

The food industry is made up of businesses that produce, manufacture, transport, and distribute food for people in the United States and throughout the world. Food production involves many activities that occur on farms and ranches, in orchards, and in fishing operations. Food manufacturing takes the raw materials harvested by producers and converts them into forms suitable for distribution and sale. The retail distribution system consists of the many food operations that store, prepare, package, serve, display, vend, or otherwise provide food for human consumption. The term **food establishment** refers to all facilities involved in food distribution.

During 2000, U.S. grocery sales grew by 3.4%, reaching $570 billion. Retail channels have changed dramatically over the last few years (Willard Bishop Consulting, 2001). There are 15 different store formats currently operating in the retail food industry. These can include:

- Conventional supermarkets
- Superstores
- Warehouse stores
- Convenience stores.

This book focuses on information related to food safety and sanitation in any retail food establishment. The term **retail food establishment** includes any store selling a line of dry retail food, canned goods or non-food items plus perishable items. A convenience store is a compact drive-to store offering a limited number of high-convenience items. The information would also apply to other types of retail food stores such as supercenters, warehouse stores, and wholesale clubs.

Food Safety-Why All the Fuss?

Everyone knows the United States has one of the safest food supplies in the world. So why all the fuss? The answer is simple. Foodborne illness happens, and it adversely affects the health of millions of Americans every year. Foodborne illness is the sickness some people experience when they eat contaminated food. It impairs performance and causes discomfort. Estimates of the number of cases of foodborne illnesses vary greatly. A recent report issued by the Centers for Disease Control and Prevention (CDC) estimates foodborne diseases cause approximately 76 million illnesses, 325,000 hospitalizations, and 5,000 deaths in the United States each year (Mead et. al., 1999).

The Food Marketing Institute (FMI) reports nearly 75% of shoppers are completely or mostly confident the food they buy in their retail food establishment is safe. Seven out of every 10 shoppers say product safety is "very important" to them when they shop for food. That is a valid reason to keep the food sold in your operation as safe as possible.

Foodborne illness costs billions of dollars each year in the form of:

- Medical expenses
- Lost work and reduced productivity by victims of the illness
- Legal fees
- Punitive damages
- Increased insurance premiums
- Lost business
- Loss of reputation for the retail food establishment.

Why Me?

You may be asking yourself, "What does all this have to do with me?" The answer is "PLENTY." Customer opinion surveys show cleanliness, price and food quality are the top three reasons people use when choosing a place to eat and shop for food. Customers expect their food to taste good and not make them sick. It is the responsibility of every retail food establishment owner, manager, and employee to prepare and serve safe and wholesome food and preserve their clients' confidence. **As a retail food manager, you must understand foodborne illness can be prevented if the basic rules of food safety are routinely followed.**

Prevention of foodborne illness must be a goal in every retail food establishment.

The CDC reports the mishandling that causes most foodborne disease outbreaks occurs within retail food operations (restaurants, retail food establishments, schools, churches, camps, institutions, and vending locations) where foods are prepared, served, and sold to the public. These foods may be eaten at the retail food establishment or sold for preparation and consumption elsewhere.

Most Cases of Foodborne Illness in Retail Food Establishments are Caused by Foods that have been:

- Improperly cooked and/or held at improper temperatures
- Handled by infected food employees who practice poor personal hygiene
- Exposed to disease-causing agents by cross contamination
- In contact with contaminated equipment not properly cleaned and sanitized
- Obtained from unsafe sources.

Changing Trends in Food Consumption and Choices

Due to changes in our eating habits and more knowledge about food safety hazards, recommendations for safe food handling are always changing. For example, retail food establishments used to cook ground beef to an internal temperature of 140°F (60°C). But that was before Shiga toxin-producing *Escherichia coli* bacteria came on the scene. Now retail food establishments are required to cook ground beef to an internal temperature of 155°F (68°C) for 15 seconds. This higher temperature is needed to destroy the Shiga toxin-producing *Escherichia coli* bacteria that may be present in the raw meat. The safety of unpasteurized juices, like apple cider and fresh-squeezed orange juice, was not a concern in the past. Now we read about outbreaks in these juices involving *Salmonella, E. coli* and parasites. As a result, most juice products are now heat-pasteurized to improve their safety.

Some emerging food safety concerns include *Listeria monocytogenes* in ready-to-eat processed foods, Shiga toxin-producing *Escherichia coli* in raw meat and unprocessed fruit juices, Hepatitis A virus in deli sandwich operations and shellfish, and parasites in fresh produce. The future may bring other types of problems.

Technologies used in the food industry are also changing. A lot of research is being done on irradiation and other types of food-processing methods that do not use heat. They may become more common in the future. You need to keep up with this information as it relates to retail food establishment operations.

Customers have less time to prepare food because more of them are working outside the home. As a result, they are buying more ready-to-eat food or products that require minimal preparation in the home. These foods are produced using a variety of processing, holding, and serving methods that help protect them from contamination.

More than 50% of the food eaten by Americans is prepared in food-processing plants, restaurants, retail food establishments, delicatessens, cafeterias, institutions, and other sites outside the home.

The Problem: Foodborne Illness

Foodborne illness is a disease caused by the consumption of contaminated food. A **foodborne disease outbreak** is defined as an incident in which two or more people experience a similar illness after eating a common food. Recent outbreaks of foodborne illness have been caused by:

- Shiga toxin-producing *Escherichia coli* bacteria in lettuce, unpasteurized apple cider, and radish sprouts
- *Salmonella spp.* in cut melons, alfalfa sprouts, ice cream, and dry cereal
- Hepatitis A virus in raw and lightly cooked oysters
- *Listeria monocytogenes* in hot dogs and luncheon meats.

These Groups of People are Highly Susceptible to Foodborne Illness:

- The very young
- The elderly
- Pregnant or lactating women
- People with impaired immune systems due to cancer, AIDS, HIV, diabetes, or medications that suppress response to infection.

Foodborne illness can cause severe reactions, even death, for individuals in these highly susceptible categories. The availability of a safe food supply is critical to these people.

Contamination

Contamination is the presence of substances or conditions in the food that can be harmful to humans. Foods can become contaminated at a variety of points as the food flows from the farm to the table. Raw foods can be contaminated at the farm, ranch, or on board a commercial fishing boat. Contamination can also occur as foods are handled during processing and distribution. Measures to prevent and control contamination must begin when food is harvested and continue until the food is consumed.

Foods can become contaminated at several points between the farm and the table.

Soil, water, air, plants, animals, and humans are some of the more common sources of contamination. Contaminants present an "invisible challenge" because you can't see them with the naked eye. Many types of food contamination can cause illness without changing the appearance, odor, or taste of food.

Contaminants can be transferred from one food item to another by cross contamination. This typically happens when microbes from a raw food are transferred to a cooked or ready-to-eat food by contaminated hands, equipment, or utensils.

Sources of contamination

Microorganisms (Germs or Microbes)

Microorganisms (also called germs or microbes) are the most common types of food contamination. Microorganisms include bacteria, viruses, parasites, and fungi that are so small they can only be seen with the aid of a microscope. Bacteria and viruses pose the greatest safety challenges in retail food establishments. Microbes are everywhere around us–in soil, water, air, and in and on plants and animals (including humans).

Most microorganisms are harmless. However, some microbes can cause problems when they get into food. The microbes that must be controlled in a retail food establishment are the ones that cause foodborne illness and food spoilage. It is important to remember the germs that cause foodborne illnesses usually do not alter the taste, odor, and appearance of the food.

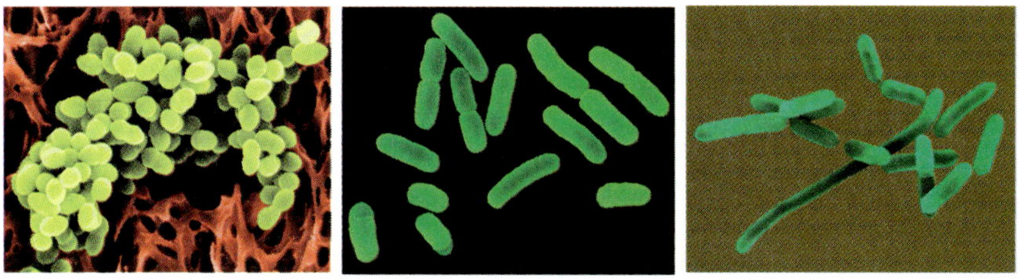

Microorganisms–common causes of foodborne illness (Copyright Dennis Kunkel Microscopy, Inc.)

The Food Flow

The food flow consists of food products and the ingredients used to make them, as they flow through a retail food establishment. The food flow begins with the purchase of safe and wholesome ingredients from approved sources. Once the food is delivered, it then flows through receiving into storage. The final stages in the food flow are preparation, display, and sale. Preparation and display include all the activities that occur between storage and consumption of the food by your customers.

You must learn to recognize foods that could have been contaminated prior to delivery.

Preparation steps frequently involve thawing, cooking, cooling, reheating, hot-holding, cold-holding, handling, and display. Improper food handling and display can lead to foodborne illness. For potentially hazardous foods (those that support bacterial growth), time and temperature must be monitored and controlled. Handling of ready-to-eat foods (those not requiring cooking) can also involve a lot of contact with an employee's hands or with food-contact surfaces. It is always important for food employees to practice good personal hygiene and use proper handwashing techniques. However, when handling and displaying ready-to-eat foods, it is crucial!

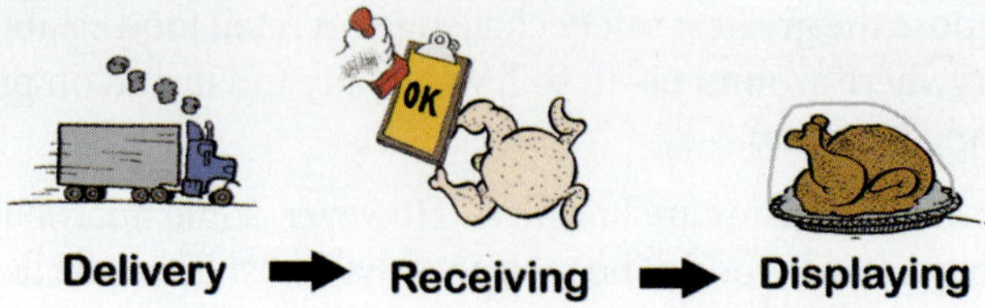

Some of the steps involved in the flow of food

Since handling and display are typically the last steps prior to consumption of the food, efficient monitoring and control of safe food practices are also critical. As products flow through a retail food establishment, they are handled many times and in many different ways. Controlling food safety and quality is of utmost importance.

A New Approach to an Old Problem

Food industry professionals and regulatory officials agree better ways to protect people from foodborne illness must be found. The *FDA Food Code* recommends using the Hazard Analysis Critical Control Point (HACCP) system as an important component of your food safety management program. The Pillsbury Company developed this innovative system in the 1960s for the NASA space program. Officials in the space program realized a foodborne illness in space could present a life-threatening situation. Therefore, it was critical all food used by the astronauts be as near 100% safe as possible.

The HACCP system follows the flow of food through the retail food establishment and identifies each step in the process where contamination might cause the food to become unsafe. When a problem step is identified, action is taken to make the product safe or, if that is not feasible, the food may have to be discarded. The HACCP system should be designed to accommodate the types of products served and the production equipment and processes used in the establishment. A more detailed discussion of the HACCP system as a component of your total food safety management program is presented in Chapter 10 of this book.

Retail food stores can control the flow of food by setting up a food safety management program that includes a HACCP system.

Facility Planning and Design

A well-planned facility with a suitable layout is essential for the smooth operation of any retail food establishment. Layout, design, and facilities planning directly influence:

- Employee safety and productivity
- Labor and energy costs
- Customer satisfaction.

Some retail food establishments are located in buildings designed and constructed specifically for a food operation. Others are located in buildings converted to accommodate food preparation, handling, display and sales. Either way, the better your facility is planned, the easier it will be to achieve your food safety goals and earn a profit.

Keeping It Clean and Sanitary

It is the responsibility of every person working in the food industry to keep things clean and sanitary. Effective cleaning of equipment reduces the chances of food contamination during preparation, storage, display, and service. **Cleaning** involves removal of visible soil from the surfaces of equipment and utensils. **Sanitary** means healthful or hygienic. It involves reducing the number of disease-causing microorganisms on the surface of equipment and utensils to acceptable public health levels. Something that is sanitary poses little or no risk to human health. Good sanitation also minimizes attraction of pests, increases the length of time equipment is in service, improves employee morale and efficiency, and is important from other aesthetic considerations.

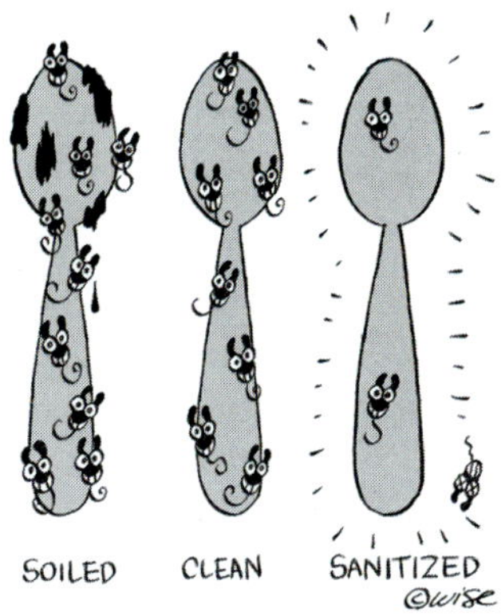

Sanitary makes sure it's safe.

Shopper satisfaction is highest in food establishments that are clean and bright and where quality food products are safely handled and displayed.

Accident Prevention and Crisis Management

Accident prevention programs are necessary in every retail food establishment. The cost of accidents may mean the difference between profit and loss. Beyond financial responsibilities, the loss of an employee's skills can disrupt operations and cause additional stress on other employees. Assuring a safe environment for employees and customers requires continuous monitoring, but the rewards make it worth the effort.

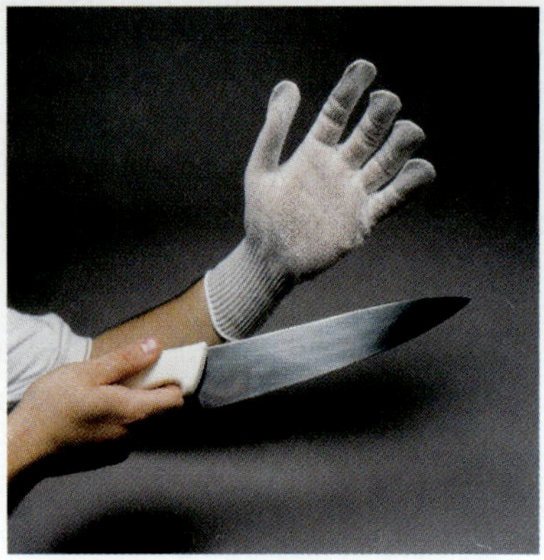

Prevention is the key to avoiding accidents.

Human error will always be a factor in retail food establishments, but training and proper equipment can help employees avoid accidents. Oven doors left open on deck ovens, poorly stacked boxes in storage areas, spilled food, and many other situations can lead to serious injuries. For example, a fall on a wet floor can cause expenses running into the thousands of dollars for lost time, employee compensation, and medical care.

In retail food establishments, another kind of crisis occurs when water supplies, electricity, or sewer systems are disrupted. The *FDA Food Code* contains specific instructions on how to operate when basic services to your establishment are lost. Public health departments can also offer assistance when you are not sure how to proceed.

Storms, tornados, hurricanes, or floods cause conditions that require changes to maintain food safety. A good disaster plan is invaluable in times of need. Managers and supervisors are expected to know how to handle these and other emergencies when they occur.

Education and Training Are Key to Food Safety

You and your employees must know the correct way to manage food safety and sanitation. The importance of teaching employees about food safety is increased by the global nature of our food supply. Control of factors during growth, harvest, and shipping is not always possible when food is produced in so many different parts of the world. Also, an error in time and temperature management, cross contamination, or personal health and hygiene of food employees can increase the risk of foodborne illness. Storage, preparation, holding, display and handling procedures are critical in the prevention of foodborne illness. Employees do not typically come to the job knowing this information. They have to be trained.

Good training can prevent foodborne illness.

The prevention of foodborne illness begins with the knowledge of where contaminants come from, how they get into food, and what can be done to control or eliminate them.

The Role of Government in Food Safety

The purpose of government regulation in food safety is to oversee the food-producing system and protect food intended for human consumption. Governmental agencies enforce laws and rules to protect food against adulteration and contamination. Regulatory personnel monitor both the process and the product to assure the safety of the food we eat.

There are several federal regulatory agencies, such as the U.S. Food and Drug Administration (FDA) and U.S. Department of Agriculture (USDA), that set food safety standards to make our food supply safer. The federal agencies are usually involved with assuring the safety of foods that are processed or prepared and then transported by interstate commerce. Most retail food establishments are governed by state or local agencies such as a health department or departments of agriculture. These agencies are likely to have the most direct impact on the everyday activities of retail food establishments.

The *FDA Food Code*

2001 ***FDA Food Code*** **book**

In 1993, the U.S. Food and Drug Administration (FDA) published the *Food Code* which replaced three earlier model codes—The Food Service Sanitation Manual Including a *Model Food Service Sanitation Ordinance of 1976*, The Vending of Food and Beverages Including a *Model Sanitation Ordinance of 1978*, and *The Retail Food Store Sanitation Code of 1982*—with one code. The FDA has made a commitment to revise the *Food Code* every four years, with the assistance of experts from state and local government, industry, professional associations, and colleges and universities. It was determined a biannual revision was necessary to keep pace with changes occurring in food safety and the food industry. In this text, the *2001 Food Code* will serve as the resource from which safe food-handling recommendations are taken and will be referred to as the ***FDA Food Code*** from this point forward.

The *FDA Food Code* is not a law. Rather, it is a set of recommendations designed for use as a model by state and local jurisdictions when formulating their own rules and regulations. You will need to obtain a copy of the health rules and regulations that apply to the retail food establishments in your jurisdiction.

The Role of the Food Industry in Retail Food Safety

The retail food industry is assuming greater responsibility for overseeing the safety of its own processes and products. Customers expect and deserve food is safe to eat. If a retail food establishment is involved in a foodborne disease outbreak, customers may retaliate by taking their business elsewhere or by seeking legal action. Financial loss and damaged reputation are some of the outcomes of a foodborne disease outbreak that can cause serious harm to the establishment found responsible for the problem. One means of preventing the harmful effects of a foodborne disease outbreak is to start a food safety management program in the retail food establishment. This helps assure proper safeguards are used during food production, handling, and display. The ability to prove a food safety system was in place at the time a foodborne disease outbreak occurred is very important. It has been deemed an acceptable defense in court cases where victims of foodborne illness have sought punitive damages.

Food Protection Manager Certification

Several organizations including the Conference for Food Protection (CFP), the FDA, and the trade associations for food retailers, such as the Food Marketing Institute (FMI), have recommended food manager certification as a means of assuring individuals who are in charge of retail food establishments are knowledgeable about food safety.

A CERTIFIED FOOD PROTECTION MANAGER IS A PERSON WHO:
• Is responsible for identifying hazards in the day-to-day operation of a retail food establishment that prepares, packages, serves, vends or otherwise provides food for human consumption • Develops or implements specific policies, procedures, or standards aimed at preventing foodborne illness • Coordinates training, supervises or directs food preparation activities and takes corrective action as needed to protect the health of the customer • Conducts in-house, self-inspection of daily operations on a periodic basis to see that policies and procedures concerning food safety are being followed.
A CERTIFIED FOOD PROTECTION MANAGER MUST DEMONSTRATE KNOWLEDGE AND SKILLS IN FOOD PROTECTION MANAGEMENT INCLUDING THE FOLLOWING AREAS:
• Identifying foodborne illness • Describing the relationship between time and temperature and the growth of microorganisms that cause foodborne illness • Describing the relationship between personal hygiene and food safety • Describing methods for preventing food contamination from purchasing and receiving • Recognizing problems and potential solutions associated with facility, equipment, and layout in a retail food establishment • Recognizing problems and solutions associated with temperature control, preventing cross contamination, housekeeping and maintenance.
A complete list of knowledge areas is presented in Appendix F.

Some state and municipal agencies already require certification of food managers, and others are considering it. Most certification programs require managers to pass a food safety examination recognized by the Conference for Food Protection. These examinations are designed to link food safety theory to practice. By passing a certification examination, the retail food manager is able to demonstrate proficiency in food protection management. Information about the Conference for Food Protection and its program for recognizing food protection manager certification examinations can be found at the organization's website at http://www.foodprotect.org

Some managers will voluntarily seek training before they take a certification examination. Some jurisdictions will require training before testing while others will require only the test. Training can involve participating in a formal course, or it could involve using a nontraditional approach such as home study, correspondence, distance learning, and interactive computer programming.

This book is written as a teaching tool to be used in conjunction with traditional and nontraditional food manager certification programs. Like the manager certification examinations, it links theory to practice. The authors of this book recognize knowing about food safety is not enough. You must also be able to apply the principles and practices that enhance food safety during preparation, display and handling.

Check it Out

Back to the Story… *Food safety is in the news almost every day. In the event of a food recall or associated foodborne illnesses, you can be sure the media will cover the story. In the food recall presented at the beginning of this chapter, ground beef products were being recalled because they contained* E. coli *O157:H7 bacteria. These organisms are a type of Shiga-toxin producing bacteria that has been shown to cause serious health effects and even death. Because of the dangers associated with these bacteria, it is considered an "adulterant" in raw ground beef, meaning it is unsafe and illegal to sell.*

The meat was most likely contaminated on the farm or during the processing into ground meat. However, the supermarket chain must be able to respond. The responsibility of the supermarket chain is to quickly remove the products from the retail shelves and communicate the risks of the recalled product to customers that may have already purchased the meat. They can do so with a good food safety management program in place and with the use of the lot codes of the XYZ meat processing company. The recall may also bring devastating economic effects. Certainly the XYZ meat processing company will be affected with the cost of 24 million pounds of beef recalled. The supermarket chain may also choose to purchase meat from another processor in the future. Customer perception may also be influenced negatively. One might guess customer buying patterns for raw and frozen ground beef may be reduced, at least for this supermarket chain in the short term.

Elements of food safety

- ✓ Everyone who works in a retail food establishment must understand the importance of food safety. It is the duty of every food establishment operator, manager, and employee to handle foods safely. Failure to do so can have a serious financial impact on your establishment and may cost you your job.
- ✓ You can protect the health and safety of your customers by developing and implementing effective food safety and sanitation practices within your establishment. In the following chapters, you will learn more about how food is contaminated and what actions are needed to prevent, control and eliminate the agents that frequently cause foodborne illness and spoilage.

Quiz 1.1 (Multiple Choice)

Please choose the BEST answer to the questions.

1. According to the Centers for Disease Control and Prevention (CDC), the estimated number of deaths due to foodborne illness each year in the United States is:
   a. 5,000.
   b. 25,000.
   c. 1,000,000.
   d. 75,000,000.
2. Sources of contamination for the food we eat include:
   a. Insects and rodents.
   b. Uncooked poultry and meat.
   c. Dirty hands and bacteria.
   d. All of the above.
3. The food safety management system being used to monitor and control contamination in potentially hazardous foods is called:
   a. integrated pest management.
   b. Cleaning and sanitizing.
   c. Hazard Analysis Critical Control Point.
   d. *FDA Food Code.*
4. Which of the following groups of people are not considered to be highly susceptible to foodborne illness?
   a. Young children.
   b. Elderly people.
   c. People with weakened immune systems.
   d. College students who live in dormitories.
5. The cost of foodborne illness can occur in the form of:
   a. Loss of sales.
   b. Medical expenses.
   c. Legal fees and fines.
   d. All of the above.

Answers to the multiple-choice questions are provided in Appendix A.

References/Suggested Readings

Centers for Disease Control and Prevention (2000). *Surveillance for Foodborne Disease Outbreaks—United States, 1993-1997.* U.S. Department of Health and Human Services. March 17, 2000. Atlanta, GA.

Food and Drug Administration (2001). *2001 Food Code.* U.S. Public Health Service. Washington, DC.

Jay, James M. (2000). *Modern Food Microbiology (6th Edition).* Aspen Publications. Gaithersburg, MD.

Mead, Paul S., Laurence Slutsker, Vance Dietz, Linda F. McCaig, Joseph S. Bresee, Craig Shapiro, Patricia M. Griffin, and Robert V. Tauxe (1999). *"Food-Related Illness and Death in the United States." Emerging Infectious Diseases,* Vol. 5, No. 5, September-October, 1999. Centers for Disease Control and Prevention. Atlanta, GA.

Potter, N., and J. Hotchkiss (1995). *Food Science (3rd Edition).* Aspen Publishers. Gaithersburg, MD.

Willard Bishop Consulting, 2001.

Learn How To:

- List the three main categories of foodborne hazards.
- Identify the difference between infections, intoxications, and toxin-mediated infections as classes of foodborne illness.
- List the factors that promote the growth of disease-causing bacteria.
- Explain how temperatures in the danger zone between 41°F (5°C) and 135°F (57°C) can affect bacterial growth.
- List the major types of potentially hazardous foods.
- Identify the characteristics common to potentially hazardous foods.

CHAPTER 2

Hazards to Food Safety

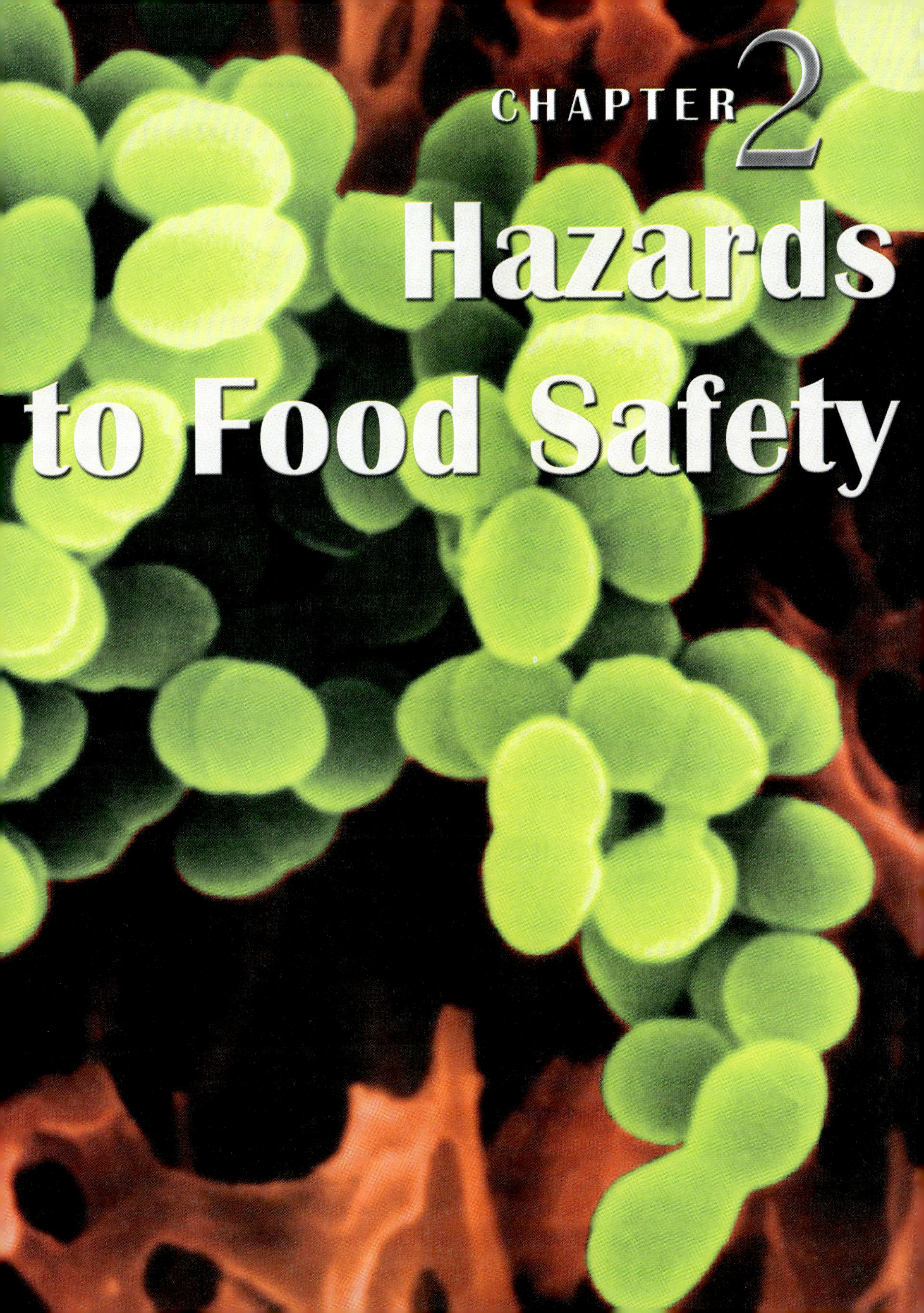

Melons cause illness

Seven customers who purchased prepared packages of cut-up cantaloupe suffered symptoms of Salmonella *infection. This included nausea, fever, vomiting, abdominal cramps, and diarrhea within 24-36 hours after eating the melon. The local health department began an investigation to discover the source of the infection. Cut-up cantaloupe from the Green Grocer Market was the only common food all had consumed.*

When the remaining packages of cantaloupe were checked, it was discovered all the melons had been contaminated with Salmonella Poona. *The melons were imported from Mexico. Many cases of* Salmonella *infection were found in at least four other states, all related to the same Mexican supplier.*

What could be done to prevent this type problem?

Foodborne Illness

Many people have had foodborne illness and not even known it. The symptoms of foodborne illness are very similar to those associated with the flu. The type of microbe, how much contamination is in the food, and the general condition of the affected person determines the severity of the symptoms.

Ingredients for foodborne illness

General symptoms of foodborne illness usually include one or more of the following:

- Headache
- Nausea
- Vomiting
- Dehydration
- Abdominal pain
- Diarrhea
- Fatigue
- Fever.

Foodborne illness is generally classified as a foodborne infection, intoxication, or toxin-mediated infection. Your awareness of how different microbes cause foodborne illness will help you understand how they contaminate food.

Classifications of Foodborne Illness	
Infection	Caused by eating food that contains living, disease-causing microorganisms.
Intoxication	Caused by eating food that contains a harmful chemical or toxin produced by bacteria or other source.
Toxin-mediated Infection	Caused by eating a food that contains harmful microorganisms that produce a toxin once inside the human intestinal tract.

Foodborne illnesses have different onset times. The **onset time** is the period between the time a person eats contaminated food and when they show the first symptoms of the disease.

Onset times vary depending on factors such as the victim's:

- Age
- Health status
- Body weight
- Amount of contaminant ingested with the food.

Anyone can become ill from eating contaminated foods. In most cases, healthy adults will have flu-like symptoms and recover in a few days. The risks and dangers associated with foodborne illness are much more serious for people who are part of highly susceptible populations. Highly susceptible populations include infants and young children, the elderly, pregnant women, individuals with suppressed immune systems due to acquired immune deficiency syndrome (AIDS), cancer, diabetes, and people taking certain types of medications. For these individuals, the symptoms and duration of foodborne illness can be much more severe—even life threatening.

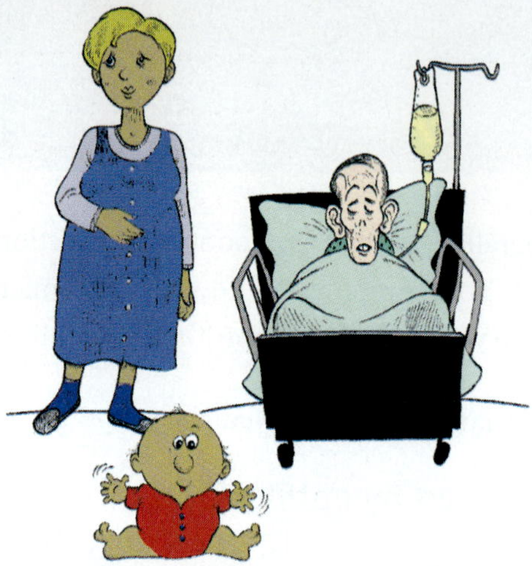

Highly susceptible populations

Foodborne Hazards

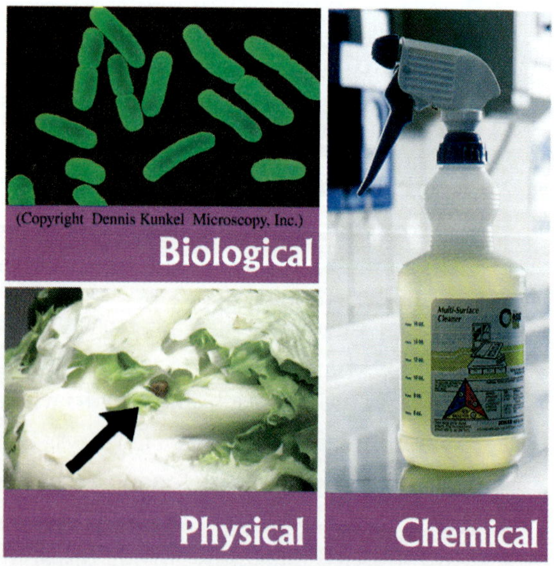

Biological hazards are by far the most important foodborne hazard in any type of retail food establishment.

A **foodborne hazard** is a biological, chemical, or physical hazard that can cause illness or injury when consumed along with the food.

Biological hazards include bacteria, viruses, parasites, and fungi, and are:

- Very small and can only be seen with the aid of a microscope
- Commonly associated with live animals, humans and with raw food products

- The most common cause of foodborne illness
- The primary target of a retail food safety program.

Chemical hazards are toxic substances that may occur naturally or may be added during the processing of food. Examples of chemical contaminants include agricultural chemicals (i.e., pesticides, fertilizers, antibiotics), cleaning compounds, heavy metals (lead and mercury), food additives, and food allergens for allergen-sensitive people. Harmful chemicals have been associated with severe poisonings and allergic reactions. Chemicals and other non-food items should be labeled clearly and never placed near food items.

Physical hazards are hard or soft foreign objects in food that can cause illness and injury. They include items such as fragments of glass, metal, unfrilled toothpicks, jewelry, adhesive bandages, and human hair. These hazards result from accidental contamination and poor food-handling practices that can occur at many points in the food chain from the farm to the customer.

Bacteria

Bacteria are single-celled microorganisms that require food, moisture, and specific temperatures to multiply. Bacteria can cause foodborne infections, intoxications, and toxin-mediated infections. In retail food establishments, most bacteria are destroyed or controlled by:

- Monitoring time and temperature
- Good personal hygiene practices
- An effective cleaning and sanitation program
- Measures that minimize cross contamination.

All bacteria exist in a "vegetative state." Vegetative cells grow, reproduce, and produce wastes just like other living organisms. Some bacteria have the ability to form structures called "spores." Spores help bacteria survive when their environment is too hot, cold, dry, acidic, or when there is not enough food. Spores are not able to grow or reproduce. More information about bacteria spores are discussed in greater detail later in this chapter.

Destroy bacterial spores or keep them from changing into the dangerous vegetative state where they can grow and cause illness.

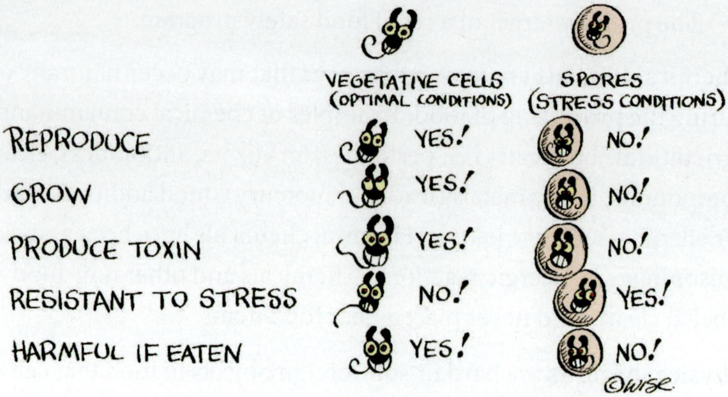

The vegetative and spore states of bacteria cells

However, when conditions become suitable for growth, a spore can germinate much like a seed. The bacterial spore can then return to the vegetative state and begin to grow again. Bacteria can survive for many months as spores and it is much harder to destroy bacteria when they are in a spore form.

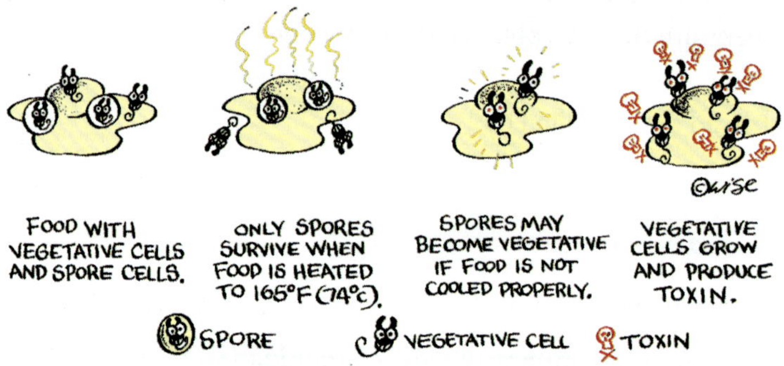

Spores and toxins pose a special challenge.

Spoilage and Disease-Causing Bacteria

Bacteria are classified as either spoilage or pathogenic (disease-causing) microorganisms.

Spoilage bacteria break down foods so they look, taste, and smell bad. They reduce the quality of food to unacceptable levels. **Pathogenic bacteria** are disease-causing microorganisms that can make people ill if the vegetative bacterial cells or their toxins are consumed with food. Both spoilage and pathogenic bacteria must be controlled in retail food establishments.

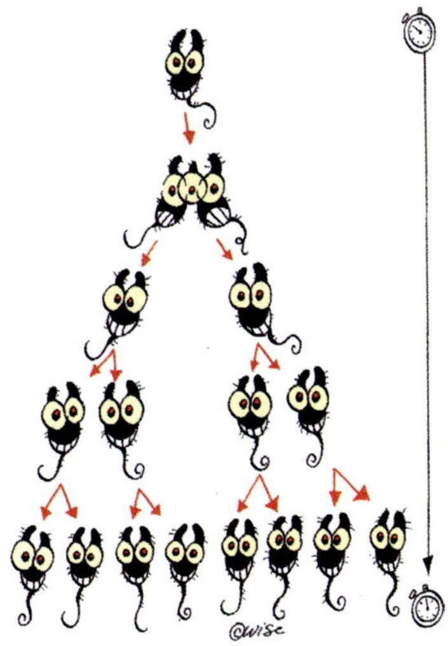

Division of bacterial cells

Bacterial Growth

Bacteria reproduce when one bacterial cell divides to form two new cells. This process is called **binary fission**. The reproduction of bacteria and an increase in the number of organisms is referred to as bacterial growth. Bacterial growth follows a regular pattern that consists of four phases:

1. **Lag phase** – Bacteria exhibit little or no growth as they adjust to their environment. This phase lasts only a few hours at room temperature but can be increased by keeping foods out of the temperature danger zone.
2. **Log phase** – Bacteria double in number every 15-30 minutes.
3. **Stationary phase** – The number of bacteria is steady as the number of new organisms being produced is equal to the number of organisms that are dying.
4. **Death phase** – Bacteria die off rapidly because they lack nutrients and are poisoned by their own wastes.

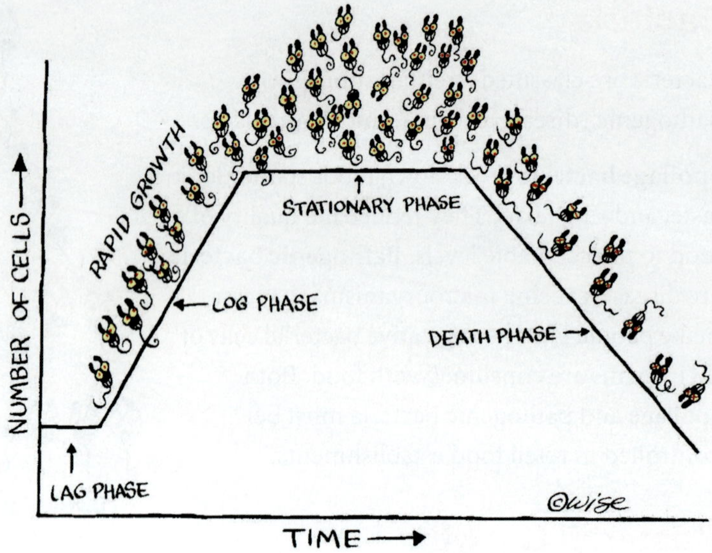

Bacterial growth curve

What Disease-causing Bacteria Need in Order to Multiply

Disease-causing bacteria need six conditions in order to multiply.

- Food
- Acid
- Temperature
- Time
- Oxygen
- Moisture.

An easy way to remember the requirements for bacterial growth is by using the acronym **F-A-T-T-O-M.**

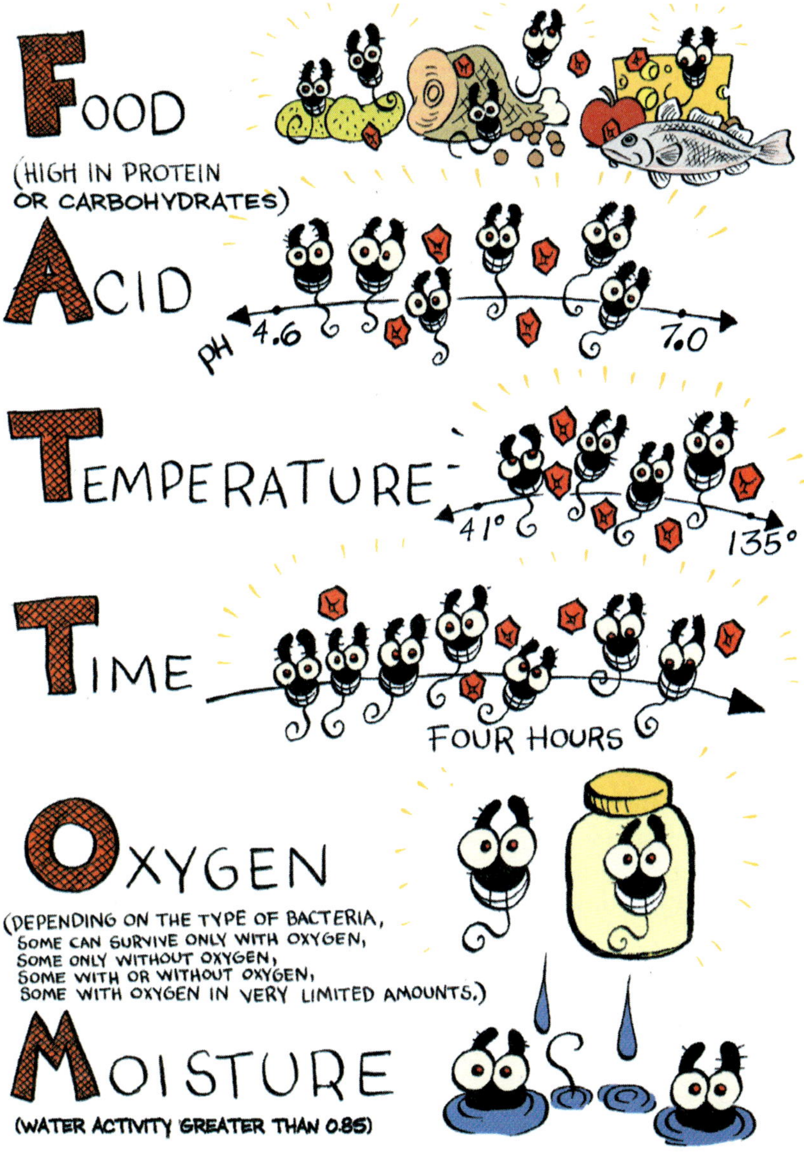

FATTOM

Since many foods naturally contain microorganisms, it is necessary to control one or more of these six conditions to prevent bacteria from multiplying.

Food

A suitable food supply is the most important condition needed for bacterial growth. Most bacteria prefer foods high in protein or carbohydrates like meats, poultry, seafood, dairy products, and cooked rice, beans, and potatoes.

Microbes eat the same foods we do.

Acidity

The **pH** symbol is used to designate the level of acidity or alkalinity of a food. You measure pH on a scale that ranges from 0 to 14.

Most foods are acidic and have a pH less than 7.0. Foods highly **acidic** (pH below 4.6), like lemons, limes, and tomatoes, will not normally support the growth of disease-causing bacteria. Pickling fruits and vegetables preserves the food by adding acids such as vinegar. This lowers the pH of the food in order to slow down the rate of bacterial growth.

A pH above 7.0 indicates the food is "**alkaline**." Only a few foods are alkaline. Examples of alkaline foods are olives, egg whites, and soda crackers.

Most bacteria that can cause foodborne illness prefer a neutral environment (pH of 7.0) but are capable of growing in foods that have a pH in the range of 4.6 to 9.0. Since most foods have a pH of less than 7.0, we have identified the range where harmful bacteria grow as 4.6 to 7.0. Many foods offered for sale in retail food establishments have a pH in this range.

Disease-causing bacteria grow best when the foods they live in and on have a pH of 4.6 to 7.0.

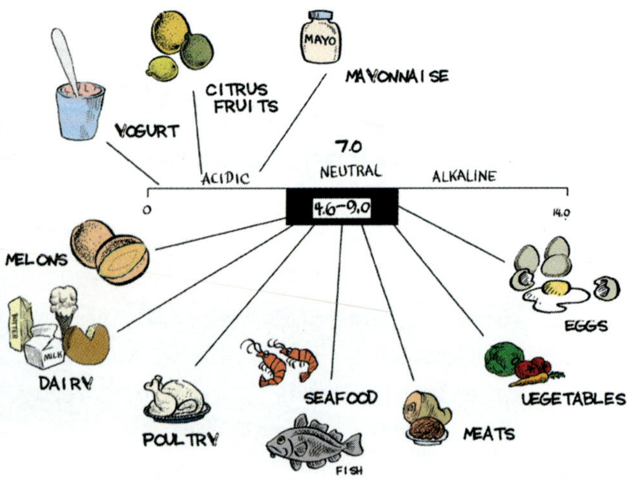

Temperature

Most disease-causing bacteria can grow within a temperature range of 41°F (5°C) to 135°F (57°C). This is commonly referred to as the food "temperature danger zone." A few disease-causing bacteria, such as *Listeria monocytogenes*, can grow at temperatures below 41°F (5°C), but the rate of growth is very slow.

Temperature abuse is the term applied to foods that have not been heated to a safe temperature or kept at the proper temperature to control growth.

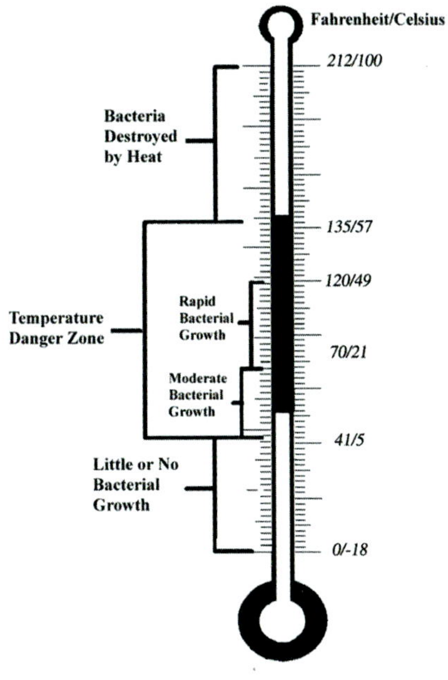

Temperature control guide

Time

For most bacteria, a single cell can generate over one million new cells in just a few hours. Because bacteria have the ability to multiply rapidly, it does not take long before many cells are produced. Bacteria need about four hours to grow to high enough numbers to cause illness. This includes the total time a food is between 41°F (5°C) and 135°F (57°C).

Time	0	15 min.	30 min.	60 min.	3 hrs.	5 hrs.
# of cells	1	2	4	16	>1000	>1 million

Bacterial cells can double in number every 15 to 30 minutes.

Careful monitoring of time and temperature is the most effective way to control the growth of pathogenic and spoilage organisms.

Oxygen

Bacteria also differ in their requirements for oxygen. **Aerobic** bacteria must have oxygen in order to grow. **Anaerobic bacteria** cannot survive when oxygen is present because it is toxic to them. These bacteria grow well in vacuum-packaged foods or canned foods where oxygen is not available. Anaerobic conditions also exist in the middle of cooked food masses such as in large stockpots, baked potatoes, or in the middle of a roast or ham.

Facultative anaerobic forms of bacteria can grow with or without oxygen.

Controlling oxygen conditions is not an effective way to prevent foodborne illness. Regardless of available oxygen, some disease-causing bacteria will be able to adapt to the conditions and grow.

Moisture

Moisture is an important factor in bacterial growth. The amount of water in a food available to support bacterial growth is called **water activity.** It is designated with the symbol A_W. Water activity is measured on a scale from 0.0 to 1.0. Water activity is a measure of the amount of water not bound to the food and is, therefore, available to support bacterial growth.

Disease-causing bacteria can only grow in foods that have a water activity higher than 0.85.

For example, fresh chicken has 60% water by volume, and its A_W is approximately 0.98. The same chicken, when frozen, still has 60% water by volume but its A_W is nearly zero. Lowering the water activity of foods to 0.85 or below preserves many foods. Drying foods or adding salt or sugar reduces the amount of available water. For example, jams and jellies that contain a lot of sugar have an A_W much less than 0.85. This alone prevents the growth of disease-causing microorganisms.

Water activity (Aw) of some foods sold in food establishments

Potentially Hazardous Foods (PHF)

The *FDA Food Code* classifies the following natural and man-made items as **potentially hazardous foods**:

- Foods of animal origin that are raw or heat-treated
- Foods of plant origin that are heat-treated or consist of raw seed sprouts
- Cut melons
- Garlic-in-oil mixtures that are not modified in a way to inhibit the growth of disease-causing microorganisms.

Potentially hazardous foods are food items that require temperature control because they are capable of supporting the rapid and progressive growth of infectious or toxin-producing microbes.

If potentially hazardous foods are held in the temperature danger zone [between 41°F (5°C) and 135°F (57°C)] for four hours or more, infectious and toxin-producing microbes can grow to dangerous levels. Potentially hazardous foods have been associated with most foodborne disease outbreaks. It is critical to control the handling and storage of potentially hazardous foods to prevent bacterial growth.

Examples of potentially hazardous foods

Ready-to-Eat Foods

Ready-to-eat foods can become contaminated if not handled properly. The *FDA Food Code* identifies the following types of foods as ready-to-eat:

- Raw animal foods that are cooked (i.e., rotisserie chicken) or frozen (i.e., sushi)
- Raw fruits and vegetables that are washed
- Fruits and vegetables that are cooked for hot-holding
- All potentially hazardous foods that are cooked and then cooled
- Bakery items such as bread, cakes, pies, fillings, or icing for which further cooking is not required for food safety
- Substances derived from plants such as spices, seasonings, and sugar
- Plant foods for which further washing, cooking, or other processing is not required for food safety, and from which rinds, peels, husks, or shells, if naturally present, are removed

> **Ready-to-eat foods are food items that are edible without washing, cooking or additional preparation by the customer or by the retail food establishment.**

- Dry, fermented sausages (i.e., dry salami or pepperoni), salt-cured meat and poultry products (i.e., prosciutto ham, country cured ham, and Parma ham), and dried meat and poultry products (i.e., jerky or beef sticks) produced in accordance with USDA guidelines and have been treated to destroy pathogens
- Thermally processed low-acid foods (i.e., smoked fish or meat) packaged in hermetically sealed containers.

Foodborne Illness Caused by Bacteria

Biological hazards are important for the retail food establishment manager to control because they lead to the majority of foodborne illness. Biological hazards are the most common agents that lead to foodborne illness.

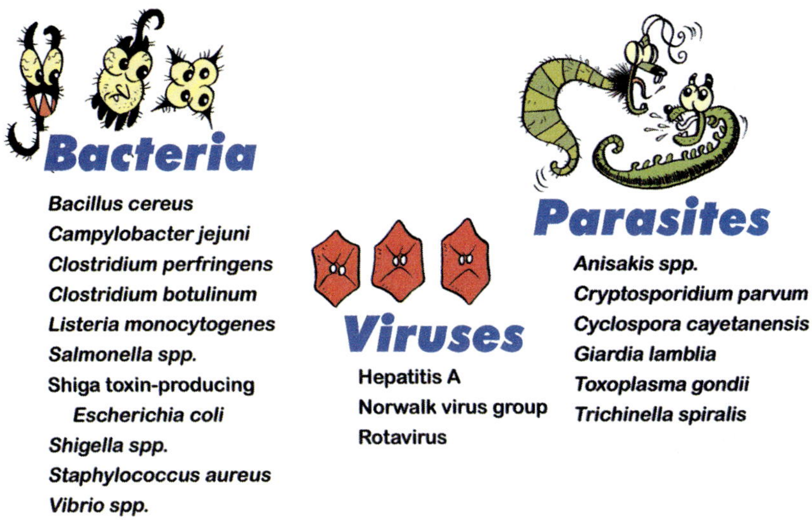

Common biological hazards in food establishments

Bacteria are classified as sporeforming and non-sporeforming organisms.

Foodborne Illness Caused by Sporeforming Bacteria

A spore structure enables a cell to survive environmental stress such as cooking, freezing, high-salt conditions, drying, and high-acid conditions.

Spores are not harmful if ingested, except in a baby's digestive system, where *Clostridium botulinum* spores can cause a disease called infant botulism. It is often recommended parents avoid serving honey to babies due to the possible presence of *Clostridium botulinum* spores. If conditions in the food are suitable for bacterial growth and the spore turns into a vegetative cell, the vegetative cell can grow in the food and cause illness if eaten.

Sporeforming bacteria are generally found in foods grown in soil, like vegetables and spices. They may also be found in animal products. They can be particularly troublesome in retail food establishments when foods are not cooled properly.

Even proper cooking or reheating may not destroy spores.

For example, a 10-gallon pot of chili was prepared for the next day's salad bar display. All the ingredients (beans, meat, spices, tomato base) were mixed together and cooked to a rapid boil. Vegetative cells should die, but spores may survive.

The chili was then stored in the 10-gallon pot and allowed to cool overnight in a walk-in refrigerator. It can take the core temperature of the chili two to three days to cool from 135°F (57°C) to 41°F (5°C)! If given enough time at the right temperature during the cooling process, sporeforming bacteria that survived the cooking process may change into vegetative cells and begin to grow.

Spores are most likely to turn into the dangerous vegetative state when:

- They are "heat-shocked" during cooking which can allow the spores to become vegetative cells
- Optimum conditions exist for growth (high protein or carbohydrates, high moisture, pH greater than 4.6)
- Temperatures are in the food temperature danger zone or between 41°F (5°C) to 135°F (57°C) for four or more hours.

It is critical hot food temperatures be maintained at 135°F (57°C) or above and cold foods should be held at 41°F (5°C) or below.

Always cook and cool foods as rapidly as possible (within four hours) to limit bacterial growth.

In the following sections, each type of biological hazard is described, the common foods and route of transmission are identified, and preventive strategies are discussed.

Foodborne Illness Caused by Non-sporeforming Bacteria

Compared to bacterial spores, vegetative cells are easily destroyed by proper cooking. There are numerous examples of non-sporeforming foodborne bacteria that are important in the food industry.

Sporeforming Bacteria

Bacillus cereus	
Causative Agent	• *Bacillus cereus*
Type of Illness	• Bacterial intoxication or toxin-mediated infection
Symptoms	• Diarrhea type: abdominal • Vomiting type: vomiting, diarrhea, abdominal cramps
Onset	• Diarrhea type: 8 to 16 hours; usually lasts 12 to 14 hours • Vomiting type: 30 minutes to 6 hours; usually lasts 30 minutes to 6 hours
Common Foods	• Diarrhea type: meats, milk, vegetables, fish • Vomiting type: rice, starchy foods, grains, cereals
Prevention	• Properly cook and hold at 135°F (57°C), cool rapidly to below 41°F (5°C), and reheat foods.

Bacillus cereus is a sporeforming bacterium that can survive with or without oxygen. It has been associated with two very different types of illnesses: one vomiting, the other diarrhea. Illness due to *Bacillus cereus* is most often attributed to foods improperly stored (cooled, hot-held), permitting the conversion of spores to vegetative cells. Vegetative cells then produce toxin in the food that leads to illness.

Properly cook and hold at 135°F (57°C), cool rapidly to below 41°F (5°C), and reheat foods.

Clostridium perfringens	
Causative Agent	• *Clostridium perfringens*
Type of Illness	• Bacterial toxin-mediated infection
Symptoms	• Intense abdominal pains and severe diarrhea
Onset	• 8 to 22 hrs
Common Foods	• Spices, gravy, improperly cooled foods (especially meats and gravy dishes)
Prevention	• Properly cook, cool, and reheat foods.

Clostridium perfringens is a nearly anaerobic (must have very little oxygen), sporeforming bacterium that causes foodborne illness. Potentially hazardous foods that have been temperature abused [not kept hot—above 135°F (57°C); or cold—below 41°F (5°C)] are frequently associated with this problem. *Clostridium perfringens* causes illness due to a toxin-mediated infection where the ingested cells colonize and then produce a toxin in the human intestinal tract. Illness due to *Clostridium perfringens* is most often attributed to foods that are temperature abused, especially those that have been improperly cooled and reheated. Foods must be cooked to 145°F (63°C) or above. Cooked foods must be cooled from 135°F (57°C) to 70°F (21°C) within two hours and from 135°F (57°C) to 41°F (5°C) in six hours. Foods must also be reheated to 165°F (74°C) within two hours and held at 135°F (57°C) until served. For quality and safety reasons, foods should be reheated only once.

Cool foods properly to prevent *Clostridium perfringens.*

Clostridium botulinum	
Causative Agent	• *Clostridium botulinum*
Type of Illness	• Bacterial intoxication
Symptoms	• Dizziness, double vision, difficulty in breathing and swallowing, headache
Onset	• 12 to 36 hours; usually lasts several days to a year
Common Foods	• Low-acid foods (pH above 4.6), which are inadequately heat-processed and then packaged anaerobically (metal can or vacuum pouch), and held in the food temperature danger zone. Examples: home-canned green beans, meats, fish, and garlic or onions stored in oil and butter respectively.
Prevention	• Properly heat-process and cool vacuum-packaged and other reduced-oxygen packaged foods, DO NOT use home-canned foods.

Clostridium botulinum is an anaerobic (must not have oxygen), sporeforming bacterium that causes foodborne intoxication due to improperly heat-processed foods, especially home-canning. Do not can foods in a retail food establishment. The organism produces a neurotoxin that is one of the most deadly biological toxins known to man. This toxin is not heat stable and can be destroyed if the food is boiled for about 20 minutes. However, botulism still occurs because people do not want to boil food that has already been cooked. Illness due to *Clostridium botulinum* is almost always attributed to ingestion of foods that were not heat-processed correctly and packaged anaerobically.

Properly heat-process and cool vacuum-packaged foods.

Non-sporeforming Bacteria

Campylobacter jejuni	
Causative Agent	• *Campylobacter jejuni*
Type of Illness	• Bacterial infection
Symptoms	• Watery, bloody diarrhea
Onset	• 2 to 5 days; usually lasts 2 to 7 days
Common Foods	• Raw poultry, raw milk, raw meat
Prevention	• Properly handle and cook raw meats and poultry, properly clean and sanitize food-contact surfaces and properly wash hands.

Campylobacter jejuni has been reported as the No. 1 cause of bacterial foodborne infection in the United States. This organism tolerates only 3 to 6% oxygen to grow. *Campylobacter jejuni* is often transferred from raw meats to other foods by cross contamination, typically from a food-contact surface (such as a cutting board or knife) or a food employee's hands.

***Campylobacter jejuni* is commonly found in raw chicken.**

Shiga toxin-producing *Escherichia coli*

Causative Agent	• Shiga toxin-producing *Escherichia coli*
Type of Illness	• Bacterial infection or toxin-mediated infection; at special risk are children up to 16 years old and the elderly
Symptoms	• Bloody diarrhea followed by kidney failure and hemolytic uremic syndrome (HUS) in severe cases
Onset	• 2 to 72 hours; usually lasts from 1 to 3 days
Common Foods	• Raw and undercooked beef and other red meats, raw finfish, improperly pasteurized milk, unpasteurized apple cider, lettuce
Prevention	• Practice good food sanitation, hand washing; properly handle and cook ground meats to an internal temperature of at least 155°F (68°C) for 15 seconds; prevent cross contamination and keep hot foods above 135°F (57°C) and cold foods below 41°F (5°C). Wash lettuce in sinks used only for food preparation. Use only pasteurized apple cider or fruit juice and milk products.

The *Escherichia coli* (or *E. coli*) group of bacteria includes four foodborne pathogens: enterotoxigenic *E. coli*, enteropathogenic *E. coli*, enterohemorrhagic *E. coli*, and enteroinvasive *E. coli*. Of particular importance is a type of enterohemorrhagic *E. coli* called Shiga toxin-producing *E. coli*. This facultative anaerobic bacteria can be found in the intestines of warm-blooded animals, especially cows. The illness caused by Shiga toxin-producing *E. coli* can be an infection or a toxin-mediated infection. Only a small amount of bacteria is required to produce an illness. A potentially hazardous food is not needed for bacterial survival. That's why apple cider, which has a pH lower than 4.6, has been implicated in cases of foodborne illness. Shiga toxin-producing *E. coli* is usually transferred to foods such as beef through contact with the intestines of slaughtered animals. Apples used for juice from orchards where cattle grazed are also suspected. Transmission can occur if employees who are carriers do not wash their soiled hands properly after going to the toilet. Cross contamination by soiled equipment and utensils may also spread Shiga toxin-producing *E. coli*.

Listeria monocytogenes	
Causative Agent	• *Listeria monocytogenes*
Type of Illness	• Bacterial infection
Symptoms	• 1) Healthy adult: flu-like symptoms. • 2) At-risk population: septicemia, meningitis, encephalitis, birth defects. • 3) Stillbirth.
Onset	• 1 day to 3 weeks; indefinite duration depending on when treatment is administered
Common Foods	• Raw meats, raw poultry, dairy products, cooked luncheon meats and hot dogs, raw vegetables, and seafood
Prevention	• Properly store and cook foods, avoid cross contamination, rotate processed refrigerated foods using FIFO to ensure timely use.

Listeria monocytogenes is a facultative anaerobic (can grow with or without oxygen) bacterium that causes foodborne infection. This microbe is important to retail food establishment operations because it has the ability to survive under many environmentally stressful conditions such as in high-salt foods and, unlike most other foodborne pathogens, can grow at refrigerated temperatures below 41°F (5°C). Transmission to foods can occur by cross contamination by people or equipment or if foods are not cooked properly.

***Listeria monocytogenes* can grow at refrigerated temperatures.**

Salmonella spp.

Causative Agent	• *Salmonella spp.*
Type of Illness	• Bacterial infection
Symptoms	• Nausea, fever, vomiting, abdominal cramps, diarrhea
Onset	• 6 to 48 hours; usually lasts 2 to 3 days
Common Foods	• Raw meats, raw poultry, eggs, milk, dairy products, pork
Prevention	• Properly cook foods; example: *Salmonella* bacteria will be destroyed when poultry is cooked to an internal temperature of 165°F (74°C) for 15 seconds and when eggs are cooked to 145°F (63°C) for 15 seconds. Clean and sanitize raw food-contact surfaces after use; make sure food employees wash their hands adequately before working with food, avoid cross contamination.

Salmonella are facultative anaerobic (grow with or without oxygen) bacteria frequently implicated as a foodborne infection. *Salmonella* are found in the intestinal tract of humans and warm-blooded animals. It frequently gets into foods as a result of fecal contamination or cross contamination. Transmission to foods is commonly through cross contamination where fecal material is transferred to food through contact with raw foods (especially poultry), contaminated food-contact surfaces (i.e., cutting boards), or infected food employees.

Eggs are a common source for *Salmonella spp.*

Shigella spp.

Causative Agent	• *Shigella spp.*
Type of Illness	• Bacterial infection
Symptoms	• Bacillary dysentery, diarrhea, fever, abdominal cramps, dehydration
Onset	• 1 to 7 days; duration depends on when treatment is administered
Common Foods	• Foods prepared with human contact: ready-to-eat salads (i.e., potato, chicken), raw vegetables, milk, dairy products, raw poultry, non-potable water, ready-to-eat meat
Prevention	• Wash hands and practice good personal hygiene, properly cook foods, avoid cross contamination, wash produce and other foods with potable water (water that is safe to drink). Do not allow individuals who have been diagnosed with shigellosis to handle food.

Shigella is facultative anaerobic bacteria that account for about 10% of foodborne illnesses in the United States. These organisms are commonly found in the intestines and feces of humans and warm-blooded animals. They cause shigellosis, a foodborne infection. The bacterium produces a toxin that causes watery diarrhea. Water contaminated by fecal material and food and utensils handled by employees who are carriers of the bacteria can cause this problem. Illness from *Shigella* is most often attributed to contaminated ready-to-eat foods handled by an infected food handler.

***Shigella* is most often attributed to foods prepared with human contact.**

Staphylococcus aureus	
Causative Agent	• *Staphylococcus aureus*
Type of Illness	• Bacterial intoxication
Symptoms	• Nausea, vomiting, abdominal cramps, headaches
Onset	• 1 to 6 hours, usually 2 to 4 hours; usually lasts 1 to 2 days
Common Foods	• Foods prepared with human contact; cooked ready-to-eat foods such as luncheon meats, ready-to-eat meat, deli salads (such as taco, potato, egg, and tuna salads), meat, poultry, custards, high-salt foods such as ham, and milk and dairy products, processed foods
Prevention	• Wash hands and practice good personal hygiene, avoid coughing and sneezing near food, do not reuse tasting spoons and ladles, properly clean and bandage cuts, burns or wounds on hands and wear plastic gloves. Cooking WILL NOT inactivate the toxin.

Staphylococcus aureus is a facultative anaerobic bacterium that produces a heat-stable toxin as it grows on foods. This bacterium can also grow on cooked, and otherwise safe, foods recontaminated by food employees who mishandle the food. *Staphylococcus aureus* bacteria do not compete well when other types of microorganisms are present. However, they grow well when alone and without competition from other microbes. These bacteria are commonly found on human skin, hands, hair, and in the nose and throat. They may also be found in burns, infected cuts and wounds, pimples, and boils. These organisms can be transferred to foods easily, and they can grow in foods that contain high salt or high sugar, and have a lower water activity. They grow well in a high-salt concentration environment, such as on hams and luncheon meats. Foods requiring considerable food preparation and handling are especially susceptible. The bacteria are also spread by droplets of saliva from talking, coughing, and sneezing near food. Food employees who improperly use tasting spoons and ladles can transfer bacteria from their mouth to food. Contaminated human hands combined with temperature abuse usually cause most problems associated with *Staphylococcus aureus*.

Vibrio spp.	
Causative Agent	• *Vibrio spp.*
Type of Illness	• Bacterial infection
Symptoms	• Headache, fever, chills, diarrhea, vomiting, severe electrolyte loss, gastroenteritis
Onset	• 2 to 48 hours
Common Foods	• Raw or improperly cooked fish and shellfish
Prevention	• Practice good sanitation, properly cool foods, implement procedures to separate raw and ready-to-eat seafood display cases, buy seafood from approved sources only.

There are three organisms within the *Vibrio* group of bacteria connected with foodborne infections. They include *Vibrio cholera, Vibrio parahaemolyticus,* and *Vibrio vulnificus.* All are important since they are very resistant to salt and are common in seafood.

Since the organism is inherent in many raw seafood, transmission to other foods by cross contamination is a concern. Most illnesses are caused by the consumption of raw or undercooked seafood.

Foodborne Illness Caused by Viruses

Viruses are now thought to be the No. 1 cause of foodborne and waterborne diseases in the United States. The viruses that cause foodborne disease differ from foodborne bacteria in several ways. Viruses are much smaller than bacteria, and they require a living host (human, animal) to replicate. Viruses do not multiply in foods. However, a susceptible person needs to consume only a few viral particles in order to experience an infection.

Viruses are usually transferred from one food to another, from a food employee to a food, or from a contaminated water supply to a food. A potentially hazardous food is not needed to support survival of viruses. The viruses of primary importance to retail food establishments are Hepatitis A and Norwalk virus. Proper hand washing and separation of raw and ready-to-eat foods are important keys to controlling the spread of foodborne viruses.

Hepatitis A

Causative Agent	• Hepatitis A
Type of Illness	• Viral infection
Symptoms	• Fever, nausea, vomiting, abdominal pain, fatigue, swelling of the liver, jaundice
Onset	• 10 to 50 days; a mild case usually lasts several weeks, more severe cases can last several months
Common Foods	• Raw and lightly cooked oysters and clams harvested from polluted waters; raw vegetables that have been irrigated or washed with polluted water; foods prepared with contact by infected employee, including salads, sliced luncheon meats, salad bar items, sandwiches, bakery products; contaminated water
Prevention	• Buy clams, oysters, and molluscan shellfish from approved sources; keep raw and ready-to-eat foods separate during storage and display; handle foods properly and cook them to recommended temperatures; wash hands and practice good personal hygiene.

Hepatitis A is a foodborne virus associated with many foodborne infections. Hepatitis A causes a liver disease called infectious hepatitis. The Hepatitis A virus is a particularly important hazard to retail food establishments because employees can harbor the virus for up to six weeks and not show symptoms of illness. Food employees are contagious for one week before onset of symptoms and two weeks after the symptoms of the disease appear. During that time, infected employees can contaminate foods and other employees by spreading fecal material from unwashed hands and nails. Hepatitis A virus is very hardy and can live for several hours in a suitable environment. The virus is transmitted by ingestion of food and water that contain the Hepatitis A virus. Raw seafood and foods handled by infected human hands are the largest threat of transmission and disease from Hepatitis A. Additional information about controls and prevention of Hepatitis A is available in the *Hepatitis A Information Guide* published by the Food Marketing Institute.

Norwalk virus	
Causative Agent	• Norwalk virus
Type of Illness	• Viral infection
Symptoms	• Vomiting, diarrhea, abdominal pain, headache, low-grade fever
Onset	• 24 to 48 hours, usually lasts 1 to 3 days
Common Foods	• Sewage-contaminated water; contaminated salad ingredients; raw clams, oysters; foods contaminated by infected food employees
Prevention	• Use potable water; cook all shellfish; handle food properly; meet time, temperature guidelines for PHF; practice good personal hygiene and wash hands and fingernails thoroughly; keep raw and ready-to-eat seafood products separate.

The Norwalk virus is another common foodborne virus associated with many foodborne infections. The virus is primarily transmitted by ingestion of food and water contaminated with feces that contain the Norwalk virus.

Foodborne Illness Caused by Parasites

Foodborne parasites are another important foodborne biological hazard. Parasites are small or microscopic creatures that need to live on or inside a living host to survive. Many parasites can enter the food system and cause foodborne illness. In this chapter, we list a few of the most troublesome ones that may appear in retail food establishments. Parasitic infection is far less common than bacterial or viral foodborne illnesses.

Anisakis spp.

Causative Agent	• *Anisakis spp.*
Type of Illness	• Parasitic infection
Symptoms	• Coughing if worms attach in throat, vomiting and abdominal pain if worms attach in stomach, sharp pain and fever if worms attach in large intestine
Onset	• 1 hour to 2 weeks
Common Foods	• Raw or undercooked seafood; especially bottom-feeding fish
Prevention	• Cook fish to the proper temperature throughout, freeze to meet *FDA Food Code* specifications, inspect seafood and handle carefully, purchase seafood from approved supplier.

Anisakis spp. are nematodes (roundworms) associated with foodborne infection from fish. The worms are about 1 to 1-1/2 inches long and the diameter of a human hair. They are beige, ivory, white, gray, brown, or pink. Other names for this parasite are "cod worm" (not to be confused with common roundworms found in cod) and "herring worm." The natural hosts of the parasite are walruses, and perhaps sea lions and otters. The worms are transferred to fish, their intermediate host, in the water in which the walruses live. Humans become the accidental host upon eating fish infested with the parasites. Humans do not make good hosts for the parasites. The worms will not complete their life cycles in humans, and eventually die.

Bottom-feeding fish such as salmon are a common source of *Anisakis spp.*

Cyclospora cayetanensis	
Causative Agent	• ***Cyclospora cayetanensis***
Type of Illness	• Parasitic infection
Symptoms	• Watery and explosive diarrhea, loss of appetite, bloating
Onset	• Usually within 1 week; symptoms persist for weeks or months if untreated
Common Foods	• Contaminated water, strawberries, raspberries, and fresh produce
Prevention	• Good sanitation and personal hygiene, purchase foods from reputable supplier.

Cyclospora cayentanensis is a parasite that has been reported much more frequently beginning in the 1990s. *Cyclospora* frequently finds its way into water and then can be transferred to foods. It can also be transferred to foods during handling. The most recent outbreaks of cyclosporiasis have been associated with fresh fruits and vegetables that were contaminated at the farm. *Cyclospora* is passed from person to person by fecal-oral transmission. Foods usually become contaminated after coming in contact with fecal material from polluted water or a contaminated food employee. The *Cyclospora* parasite may take days or weeks after a person eats a contaminated food to become infectious.

Wash berries to remove contaminants

Trichinella spiralis

Causative Agent	• *Trichinella spiralis*
Type of Illness	• Parasitic infection from a nematode worm
Symptoms	• Early symptoms: nausea, vomiting, diarrhea, sweating, abdominal pain; in later stages: fever, swelling of tissues around eyes, muscle stiffness
Onset	• 2 to 28 days; death may occur in severe cases
Common Foods	• Primarily undercooked pork products and wild game meats (bear, walrus)
Prevention	• Cook foods to the proper temperature throughout, i.e., no pink color in cooked pork products.

Trichinella spiralis is a foodborne roundworm that causes a parasitic infection. It must be eaten with the infected fleshy muscle of certain meat-eating animals to be transmitted to a new host. Meat-eating, scavenger animals frequently carry this parasite. These animals are exposed to the parasite when they eat infected tissues from other animals and garbage that contains contaminated raw-meat scraps.

Cook pork so there is no pink color inside.

Cryptosporidium parvum	
Causative Agent	• *Cryptosporidium parvum*
Type of Illness	• Parasitic infection
Symptoms	• Severe watery diarrhea
Onset	• Within 1 week of ingestion
Common Foods	• Contaminated water, food contaminated by infected food employees
Prevention	• Use potable water supply; practice good personal hygiene and hand washing.

Cryptosporidium parvum is a parasite found in water that has been contaminated with cow feces. The parasite causes foodborne infection and is considered an important source of nonbacterial diarrhea in the United States. The parasite could occur, theoretically, on any food touched by a contaminated food handler. It is primarily transmitted by a water supply contaminated with feces and by fecal contamination of food and food-contact surfaces. Parasite prevention starts with providing a potable water supply in the retail food establishment and handling foods carefully to prevent contamination and cross contamination. Food employees must practice good personal hygiene and wash hands thoroughly before working with food and after going to the toilet.

Problems Caused by Fungi

Yeast and molds make up the group called fungi. Yeasts and molds mainly contribute to food spoilage. When yeast grows in a food, it leads to undesirable characteristics of a food including production of gases, acids, and alcohol. The food may taste or smell "sour" or the package may swell. Yeasts do not lead to foodborne illness. In fact, yeasts are often used to produce fermented foods such as beer, wine, and cheeses. Molds usually grow in foods low in moisture (such as bread or cheese), acidic (fruit juices), or high in sugar (jams and jellies). They often appear very colorful or "cotton-like" or "powdery" in appearance. Like yeasts, molds themselves do not cause foodborne illness. However, if they grow long enough on foods, they can produce a substance called a "mycotoxin." Mycotoxins can cause foodborne illness and some cancers. Mycotoxins are usually considered chemical hazards.

Foodborne Illness Caused by Chemicals

Chemical hazards are usually classified as either naturally occurring or man-made chemicals. Naturally occurring chemicals include toxins produced by a biological organism. Man-made chemicals include substances added, intentionally or accidentally, to a food during processing. A summary of some of the more common naturally occurring and man-made chemicals is provided below.

Types of Chemical Hazards in a Retail Food Establishment	
Naturally Occurring:	**Man-made Chemicals:**
• Allergens	• Cleaning solutions
• Ciguatoxin	• Food additives
• Mycotoxins	• Pesticides
• Scombrotoxin	• Heavy metals.
• Shellfish toxins.	
(Source: *FDA Food Code*)	

Naturally Occurring Chemicals

Food Allergens

Between 5 and 8% of children and 1 to 2% of adults are allergic to certain chemicals in foods and food ingredients. These chemicals are commonly referred to as food allergens. A food allergen causes a person's immune system to overreact. Some common symptoms of food allergies are hives; swelling of the lips, tongue, and mouth; difficulty breathing or wheezing; and vomiting, diarrhea, and cramps. These symptoms can occur in as little as five minutes. In severe situations, a life-threatening allergic reaction called anaphylaxis can occur. Anaphylaxis is a condition that occurs when many parts of the body become involved in the allergic reaction. Symptoms of anaphylaxis include itching and hives; swelling of the throat and difficulty breathing; lowered blood pressure and unconsciousness.

The FDA has identified the "Big Eight" from the 170 different foods known to cause allergic reactions. About 90% of ingredients include the "Big Eight" and are considered to be "major serious allergens." The only way for a person who is allergic to one of these foods to keep from having an allergic reaction is to avoid the food containing the allergen. In many cases, it doesn't take much of the food to produce a severe reaction. As little as half a peanut can cause a severe reaction in highly sensitive people.

Common Food Allergens The Big Eight	
Milk	Soy
Eggs	Tree nuts
Wheat proteins	Fish
Peanuts	Shellfish.

Allergies can be very serious. You need to know which foods in your establishment contain these ingredients. The FDA requires ingredients to be listed on the label of packaged foods. Always read label information to determine if food allergens may be present.

The responsibility of retailers in the case of prepackaged foods is to assure appropriate label warnings are given to potential purchasers. Retailers have special responsibilities in regard to foods sold from open containers (i.e., on delicatessen counters). When such foods contain a major food allergen, the warning should be clearly displayed by the foods in question. Moreover, staff must be trained to take great care to avoid cross contamination such as might occur when using the same ladle or other handling equipment for a food containing a major food allergen and one that does not contain it.

Foods containing potential contaminants must be properly labeled.

Ciguatoxin	
Causative Agent	• Ciguatoxin
Type of Illness	• Fish toxin originating from toxic algae of tropical waters
Symptoms	• Vertigo, nausea, hot/cold flashes, diarrhea, vomiting, shortness of breath
Onset	• 30 minutes to 6 hours; usually lasts a few days but death can occur from concentrated dose of toxin
Common Foods	• Marine finfish including grouper, barracuda, snapper, jack, mackerel, triggerfish, reef fish
Prevention	• Purchase fish from a reputable supplier, cooking WILL NOT inactivate the toxin.

Ciguatoxin poisoning is an example of an intoxication caused by eating contaminated tropical reef fish. The toxin is found in tiny, free-swimming sea creatures called algae that live among certain coral reefs. When small reef fish eat the toxic algae, it is stored in the flesh, skin, and organs. When bigger fish such as barracuda eat the small reef fish such as mackerel, mahi mahi, bonito, jackfish, and snapper, the toxin accumulates in the flesh and skin of the consuming fish. The toxin does not affect the contaminated fish. The toxin is heat stable and not destroyed by cooking. At the present time, there is no commercially known method to determine if ciguatoxin is present in a particular fish. The toxin is transferred to finfish when they eat toxin-containing algae or other fish that contain the toxin.

Ciguatoxin can be found in marine finfish such as red snapper.

Scombrotoxin	
Causative Agent	• **Scombrotoxin**
Type of Illness	• Seafood toxin originating from histamine-producing bacteria
Symptoms	• Dizziness; burning feeling in the mouth; facial rash or hives; shortness of breath; peppery taste in mouth; headache; itching, teary eyes; runny nose
Onset	• Few minutes to half-hour; recovery usually occurs in 8 to 12 hours
Common Foods	• Tuna, mahi-mahi, bluefish, sardines, mackerel, anchovies, amberjack, abalone, Swiss cheese
Prevention	• Purchase fish from a reputable supplier; store fish between 32°F (0°C) and 39°F (4°C) to prevent growth of histamine-producing bacteria; toxin IS NOT inactivated by cooking.

Eating foods high in a chemical compound called histamine causes scombrotoxin, also called histamine poisoning. Histamine is usually produced by certain bacteria when they decompose foods containing the protein histidine. Dark meat of fish has more histidine than other fish meat. Histamine is not inactivated by cooking. Over time, bacteria inherent to a particular food can break down histidine and cause the production of histamine. Leaving fish out at room temperature usually results in histamine production.

Added Man-made Chemicals

There is an extensive list of chemicals added to foods that may pose a potential health risk. Intentionally added chemicals may include food additives, food preservatives, and pesticides. Pesticides leave residues on fruits and vegetables and can usually be removed by a vigorous washing procedure. Non-intentionally added chemicals may include contamination by chemicals such as cleaning and sanitary supplies. Also, chemicals from containers or food-contact surfaces of inferior metal that are misused may lead to heavy-metal or inferior-metal poisoning (cadmium, copper, lead, galvanized metals, etc.).

Foodborne Illness Caused by Physical Hazards

Physical hazards are foreign objects in food that can cause illness and injury. They include items such as fragments of glass, metal shavings from dull can openers, unfrilled toothpicks that may contaminate sandwiches, human hair and jewelry, or bandages that may accidentally be lost by a food handler and enter food. Stones, rocks, or wood particles may contaminate raw fruits and vegetables, rice, beans, and other grain products.

Common physical hazards in a food establishment

Physical hazards commonly result from accidental contamination and poor food-handling practices that can occur at various points in the food chain from harvest to consumer.

To prevent physical hazards:

- Wash raw fruits and vegetables thoroughly
- Visually inspect foods that cannot be washed (such as ground beef).

Food employees must be taught to handle food safely to prevent contamination by unwanted foreign objects such as glass fragments and metal shavings. Finally, food employees should not wear jewelry when involved in the production of food, except for a plain wedding band.

Problems in Other Countries Related to Food

Mad Cow Disease or bovine spongiform encephalolpathy (BSE) is a fatal brain disorder that occurs in cattle and is caused by some unknown agent. There seems to be a connection between animal feed made from the parts of sheep that carry the organisms called scrapie. Use of animal feed made from such ingredients has been banned in the United States.

The connection between BSE and humans was uncovered in Great Britain in the 1990s when several young people died of a brain disorder, a new variation of a rare problem, Creutzfeldt-Jakob disease (CJD) that typically strikes the elderly.

Efforts of the U.S. Food and Drug Administration (FDA), U.S. Department of Agriculture (USDA), the Centers for Disease Control and Prevention (CDC), and other federal organizations, including state regulatory and health agencies, have kept the diseases from occurring in this country. Advisories are regularly issued to those traveling to Great Britain and Europe about the safety of eating beef and beef products.

Foot- (hoof-) and-mouth disease is not a food safety concern, but is a concern for animal health and economics. The disease is a highly contagious viral infection of cattle, sheep, goats, deer, and other cloven-hoofed animals. It causes blisters on the mouth, teats, and soft tissues of the animal's feet and mouth. The animals rarely recover.

This disease is not a significant health concern for humans and was last found in the United States in 1929. In 2001, an outbreak of foot-and-mouth disease occurred in the United Kingdom (England, Scotland, Wales, Northern Ireland) and Europe. A large number of animals were destroyed to contain the infection.

Affected areas were placed under quarantine. The danger from this disease is to animals, not humans; therefore, foot-and-mouth disease is not classified as a foodborne illness.

Check it Out

Back to the Story... ***Melons contaminated with bacteria on the rind, even if they have been washed on the farm, may still carry the microbes. When melons are cut, the bacteria are carried through the fruit. Cantaloupe provides an ideal source of food for bacterial growth. Retail food employees and consumers alike are urged to wash and rinse all fruits and vegetables in running water just before preparing and consuming them. Receiving cantaloupes from a reputable supplier and proper in-house washing procedures may have prevented the outbreak involving*** **Salmonella Poona** ***from occurring.***

- ✓ Biological hazards such as bacteria, viruses, and parasites continue to cause problems. An increase in foodborne illness associated with fresh produce was described at the beginning of this chapter. This change is due to several factors. An increase in the number of people in the highly susceptible population exposes more individuals to foodborne illness. The number and types of hazards also seem to be on the increase. Retail food establishment managers need to be aware of the change in eating habits, marketing patterns, and emerging pathogens in our food supply.
- ✓ With fruit and vegetable products, we learned potentially hazardous foods are not the only problems. Apple cider, for example, has a pH well below 4.6, not a potentially hazardous food. However, certain microbes like Shiga toxin-producing *Escherichia coli* can survive and still cause illness.
- ✓ There are many foodborne hazards a retail food establishment may encounter. They are classified as biological, chemical, or physical hazards. These hazards differ depending on the type of food and method of preparation involved. Retail food establishments are typically toward the end of the food production chain. This is where foods are prepared or sold for consumer preparation. Therefore, it is very

important to control and prevent foodborne hazards as much as possible to reduce the risk of foodborne illness associated with your establishment. Control and prevention of foodborne hazards in a retail food establishment start with understanding the different types of foodborne hazards. The next step is to understand how to control foodborne hazards with time/temperature control, good personal hygiene, cleaning and sanitation, and prevention of cross contamination. Prevention, using these four approaches, will be the focus of Chapter Three.

Quiz 2 (Multiple Choice)

Choose the BEST answer for each question.

1. If two or more persons become ill as a result of eating the same food, it is called a foodborne:
   a. Illness.
   b. Contamination.
   c. Hazard.
   d. Disease outbreak.
2. Bacteria grow best within a narrow temperature range called the temperature danger zone. The temperature danger zone is between:
   a. 0°F (-18°C) and 220°F (104°C).
   b. 0°F (-18°C) and 135°F (57°C).
   c. 41°F (5°C) and 135°F (57°C).
   d. 41°F (5°C) and 220°F (104°C).
3. The pH of a food refers to:
   a. Available water.
   b. Acidity or alkalinity.
   c. Amount of oxygen.
   d. Amount of protein.
4. The most effective way to control the growth of bacteria in a retail food establishment is by controlling:
   a. Time and temperature.
   b. pH and oxygen conditions.
   c. Temperature and water activity.
   d. Time and food availability.

5. Which of the following is not considered a potentially hazardous food group?
   a. Red meats.
   b. Fish and shellfish.
   c. Poultry and eggs.
   d. Dried grains and spices.

References/Suggested Readings

Adams, M. R. and M. O. Moss (2000). *Food Microbiology*. Royal Society of Chemistry. London, England.

Banwart, G. J. 1989. *Basic Food Microbiology 2nd ed.* Van Nostrand Reinhold:New York.

Chin, James (Ed.) (2000). *Control Of Communicable Diseases in Man (17th ed.)*. American Public Health Association. Washington, DC.

Doyle, Michael P., Beuchat, Larry R. and Thomas J. Montville (1997). *Food Microbiology: Fundamentals and Frontiers*. American Society for Microbiology, Washington, D.C.

Food and Drug Administration (2001). *FDA 2001 Food Code*. U.S. Public Health Service. Washington, DC.

FMI Hepatitis A Information Guide, 2002.

Food Protection Report, June 2002 Vol. 18, No. 6. Pike & Fishers. Silver Springs, MD

Jay, James J. (2000). *Modern Food Microbiology 6th ed.* Aspen Publishers, Gaithersburg, MD.

Longrèe, K., and G. Armbruster (1996). *Quantity Food Sanitation*. Macmillan. New York, NY.

McSwane, D., Rue, N., Linton, R. (2004). *Essentials of Food Safety and Sanitation, 4th ed.* Prentice Hall, Upper Saddle River, NJ.

Learn How To:

- Identify potential problems related to temperature abuse of foods.
- Describe how to properly measure and maintain food temperatures to assure that foods are safe for consumption.
- Identify potential problems related to a food worker's poor personal hygiene.
- Explain how to improve personal hygiene habits to reduce the risk of foodborne illness.
- Identify potential problems related to cross contamination of food.
- Discuss procedures and methods to prevent cross contamination.

CHAPTER 3

Factors that Affect Foodborne Illness

Three Children in Critical Condition ...

Eight people became ill in a Florida community after eating boxed lunches purchased at the local supermarket. Three children were hospitalized and in serious condition for several days. A local church group picked up their 24 meals and served them within 20 minutes of purchase. Some people in the group complained of diarrhea, fever, nausea, vomiting and abdominal cramps. The children were the first and most seriously affected within 12 hours of eating their meals. There were three types of meals available—chicken salad, cut-up fruit cup, potato salad and a chocolate chip cookie; ham salad, piece of watermelon, macaroni salad and an oatmeal cookie; and a roast beef sandwich, fresh peaches, potato salad and chocolate fudge brownie. All items were prepared onsite.

The corporate office investigated the flow of food through the preparation process. The work stations for preparation were separated by a handwashing sink, with the bakery and the deli sharing equipment and utensils. The deli area was responsible for many of the ready-to-eat food preparation items such as fruit and entrée salads. The bakery provided the cookies already baked and bagged.

It was later learned that 50% of the people who became sick had the roast beef sandwich meals and 50% had the ham salad meals. Upon further questioning, it was learned the children had not eaten their sandwiches, but had eaten their fruit and most had eaten the cookie. All the adults in the group reported eating their sandwiches and potato or macaroni salad. The employees were interviewed and it was discovered the same cutting board had been used for the fruit and the raw chicken prior to cooking it for the salad. Employees in the work areas were observed to have sanitizing solution available and fresh aprons upon request. Interviews with employees revealed they had been wearing disposable gloves, but it was not clear they had been changing them properly. Upon investigation it was unclear whether the employees actually used the sanitizing solutions on the counters or the cutting boards.

Factors That Contribute to Foodborne Illness

The Centers for Disease Control and Prevention (CDC) is an agency of the federal government. One of CDC's primary responsibilities is to collect statistics about diseases that affect people in the United States, including foodborne illness. CDC statistics show most outbreaks of foodborne disease occur because *food is mishandled*. Some of the major contributors to foodborne illness are presented below.

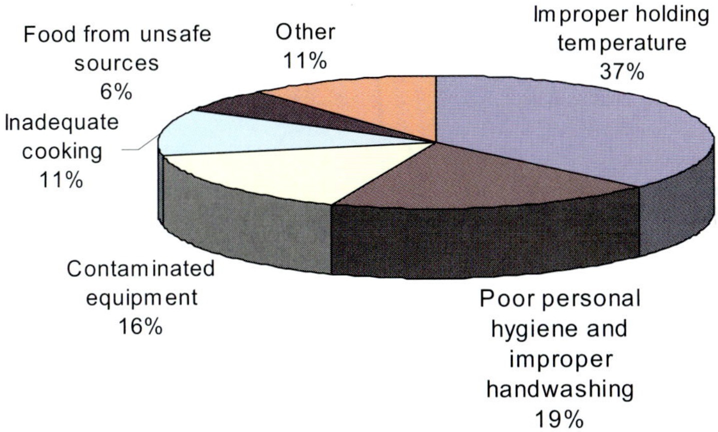

Major contributors to foodborne illness (Source: CCD-MMWR 2000)

What Is Time and Temperature Abuse?

Controlling temperature is perhaps the most critical way to assure food safety. Most cases of foodborne illness can in some way be linked to temperature abuse. The term **temperature abuse** is used to describe situations when foods are:

- Exposed to temperatures in the temperature danger zone for enough time to allow growth of harmful microorganisms
- Not cooked or reheated sufficiently to destroy harmful microorganisms.

In Chapter 2, *Hazards to Food Safety,* you learned harmful microbes can grow in potentially hazardous foods when temperatures are between 41°F (5°C) and 135°F (57°C), the temperature danger zone. Keep the internal temperatures, inside the core of a food item, out of the temperature danger zone [41°F (5°C) to 135°F (57°C)] to prevent harmful microbes from growing. Higher temperatures destroy microbes, however, toxins produced by microbes may or may not be affected by normal cooking temperatures.

The Temperature Danger Zone

Keep cold food temperatures below 41°F (5°C) and out of the temperature danger zone to prevent most microbes from growing. Bacteria that can grow at lower temperatures do so very slowly.

There are unavoidable situations during food production when foods must pass through the temperature danger zone such as:

- Cooking
- Cooling
- Reheating
- Food handling (slicing, mixing and sandwich assembling).

During these activities, you must minimize the amount of time foods are in the temperature danger zone to control microbial growth. When it is necessary for a food to pass through the temperature danger zone, do it as quickly as possible. In addition, foods should pass through the danger zone as few times as possible.

Heating foods improves texture and flavor and also destroys harmful microorganisms. As you learned in Chapter 2, many raw foods naturally contain harmful microbes or can become contaminated during handling. When you cook and reheat foods properly, microbes are reduced to safe levels or are destroyed. *Cooking and reheating are two very important processes for safe food management.*

How to Measure Food Temperatures

Maintaining safe food temperatures is an essential and effective part of food safety management. You must know *how* to measure food temperatures correctly to prevent temperature abuse. Thermometers, thermocouples, and other devices are used to measure the temperature of stored, cooked, hot-held, cold-held, and reheated foods. The following chart shows different types of thermometers and their features.

A temperature-measuring device with a small diameter probe must be available to measure the temperature of thin foods such as meat patties and fish filets.

An important rule to remember for avoiding temperature abuse is:

Keep Hot Foods Hot, Keep Cold Foods Cold, or Don't Keep the Food at All.

Thermometer	Features/Uses:
Dial-face, metal stem type (bi-metallic)  Courtesy of Cooper Instrument Corp.	• Most common type of thermometer used • Used to measure internal food temperature at every stage in the flow of food • Measures temperatures ranging from 0°F (-18°C) to 220°F (104°C) with 2°F increments • Stem of bi-metallic thermometer must be inserted at least 2 inches into the food item being measured.
Digital 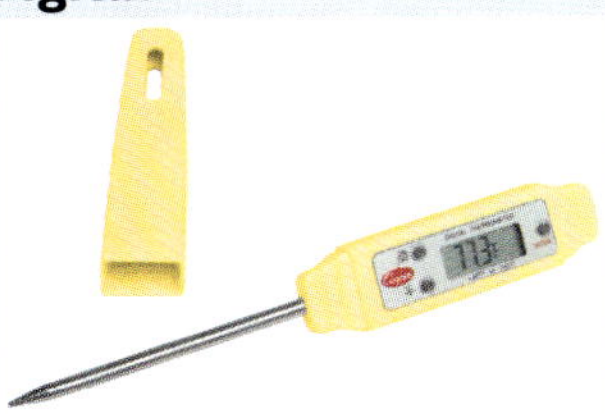 Courtesy of Cooper Instrument Corp.	• Displays the temperature numerically • Measures a wider range of temperatures than a dial-faced thermometer.

Thermocouple

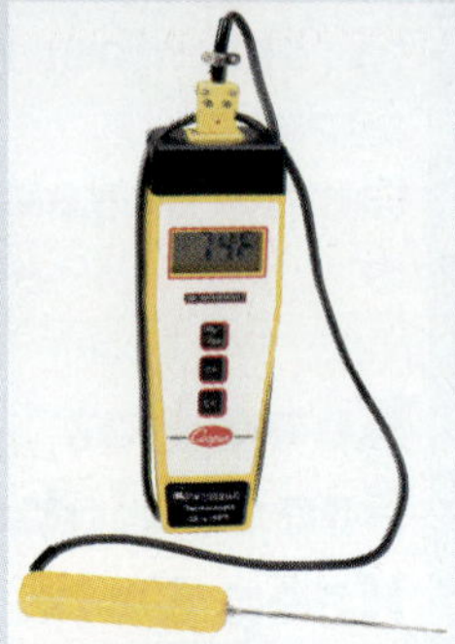

Courtesy of Cooper Instrument Corp.

- Provides a digital readout of the temperature
- Has a wide variety of interchangeable probes
- Sensing portion is often at the tip of the probe.

Infrared

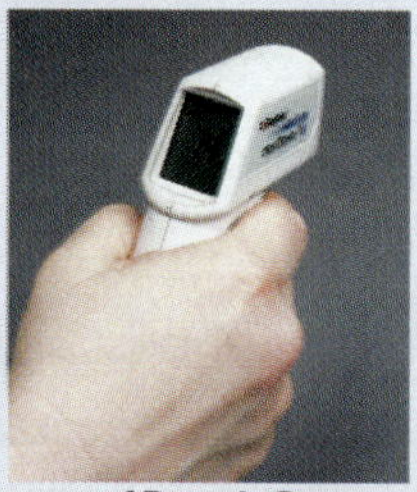

Courtesy of Raytek Corporation

- Measures the surface temperature of food without actually touching the food (reduces the chance of cross contamination)
- Requires about 20 minutes to adjust after use for hot and cold temperatures ("thermal shock") before use
- Accuracy must be checked frequently.

T-Sticks (melt devices)

Courtesy of T-Stick

- Measure only one temperature
- Change color when indicated temperature is reached
- Used to monitor food temperatures and sanitizing temperature in dishwashing machines.

Built-in

- Used to monitor air temperature in refrigerated and frozen cases.

Maximum Registering (holding)

Courtesy of Cooper Instrument Corp.

- Used to measure the temperature of hot water used to sanitize items in mechanical warewashing machines
- Becoming less popular in retail food establishments because it contains mercury (an environmental contaminant).

Thermometer Guidelines:

- Temperature-measuring devices typically measure food temperatures in degrees Farenheit (denoted as °F), degrees Celsius (denoted as °C), or both.
- Food temperature-measuring devices scaled only in Celsius or dually scaled in Celsius and Fahrenheit must be accurate to +1.8°F (+1°C). Food temperature-measuring devices scaled in Fahrenheit only must be accurate to +2°F.
- Mercury-filled and glass thermometers should not be used in retail food establishments.
- Clean and sanitize thermometers properly to avoid contaminating food that is being tested. This is very important when testing raw and then ready-to-eat food items. To clean and sanitize a food thermometer, wipe off any food particles, place the stem or probe in sanitizing solution for at least 5 seconds, then air-dry.
- When monitoring only raw foods, or only cooked foods being held at 135°F (57°C), wipe the stem of the thermometer with an alcohol swab between measurements.

When and How to Calibrate a Thermometer

Before you use a thermometer you need to calibrate it, or make sure it is working correctly. Dial-face metal stem type (bi-metal) thermometers should be calibrated:

- Before their first use
- At regular intervals
- If dropped or otherwise damaged
- If used to measure extreme temperatures
- Whenever accuracy is in question.

Calibrate dial-faced thermometers by the boiling point or ice point method. See specific directions for calibration as described in the following section. Use pliers or an open-ended wrench to adjust the indicator needle.

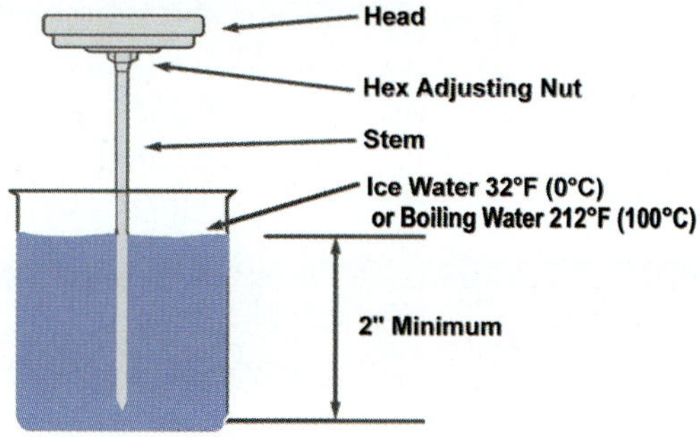

Calibrating a dial-face thermometer
(Source: Cooper Instrument Corporation)

Boiling Point Method

Immerse at least the first 2 inches of the stem from the tip (the sensing part of the probe) into boiling water and adjust the needle to 212°F (100°C). At higher altitudes, the temperature of the boiling point will vary. Consult your local health department if you have any questions about the boiling point temperature in your area.

Ice Point Method

Insert the probe into a cup of crushed ice. Add enough cold water to remove any air pockets that might remain. Wait until the temperature stabilizes and adjust the needle to 32°F (0°C).

Measuring Food Temperature

The sensing portion of a food thermometer is at the end of the stem or probe. On the bi-metal thermometer, the sensing portion extends from the tip to the "dimple" mark that is typically 1 inch up the stem. An average of the temperature is measured over this distance. The sensing portion for digital and thermocouple thermometers is closer to the tip of the probe.

The approximate temperature of packaged foods can be measured accurately without opening the package. Place the stem or probe of the thermometer between two packages of food or fold the package around the stem or probe to make good contact with the packaging.

Insert the probe at least 2" into the product.

Measure the temperature of packaged salads with an infrared thermometer.

Wait for the temperature to stabilize before removing probe.

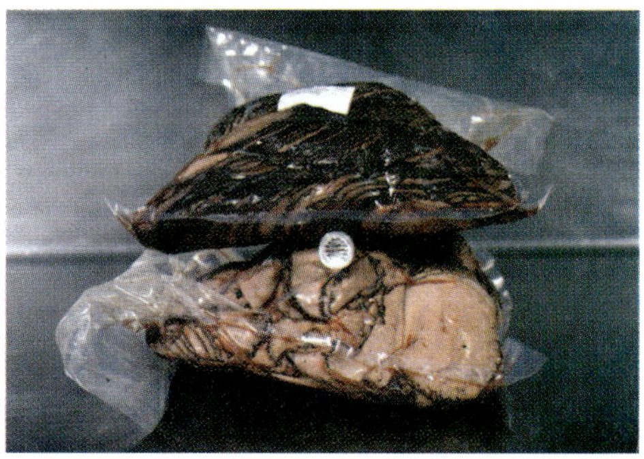

Place thermometer between packages of prepared foods to measure temperature.

How to Accurately and Safely Measure Food Temperatures:

- Use an approved temperature-measuring device that measures temperatures from 0°F(-18°C) to 220°F(104°C)
- Locate the sensing portion of the measuring device
- Calibrate the measuring device using the ice or boiling point method
- Clean and sanitize the probe of the temperature-measuring device according to procedure
- Measure the internal temperature of the food by inserting the probe into the center or thickest part of the item, at least 2 inches for a dial thermometer and 1 inch for digital thermometers
- Always wait for the temperature reading to stabilize.

Preventing Temperature Abuse

Controlling temperatures of potentially hazardous food is important in almost all stages of food handling. Measuring temperatures of potentially hazardous food is an important responsibility for all food handlers. The following chart lists safe temperature guidelines for working with food throughout the flow of food. Each of these guidelines will be discussed throughout the remainder of this book.

Time and Temperature

Receiving and Storing:

Frozen and refregerated receiving/storage practices prevent or slow the growth of harmful microorganisms

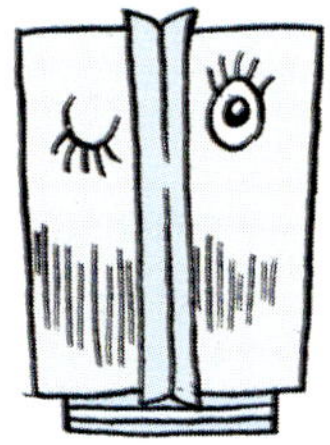

Food Product	Internal Temperature
Frozen Foods	Solidly frozen 0°F (-18°C) recommended
Refrigerated Foods	41°F (5°C) or lower
Raw Shell Eggs	45° F (7° C) or below ambient temperature

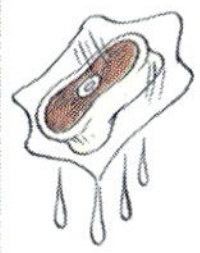

Thawing:

Take food from frozen to non-frozen to minimize the product's time in the temperature danger zone. Keep ready-to-eat foods below 41°F (5°C) at all times.

Method	Internal Temperature	Times
In refrigerator	41°F (5°C) or lower	typically takes 2-3 days
Submerged under cool running water 70°F (21°C)	Product not to exceed 41°F (5°C); Water temperature not to exceed 70°F(21°C)	4 hours or less (counts toward time in the food temperature danger zone).

Cooking:

Safely getting a food product from raw to ready-to-eat with minimum time held at internal temperature before serving

Food Product	Minimum Internal Temperature	Times
*Beef Roast (rare)	130°F (54°C) 140°F (60°C)	112 minutes 12 minutes
Beef (other than roasts), Pork (other than roasts), Fish	145°F (63°C)	15 seconds
Ground Beef, Ground Pork, Ground Game Animals	155°F (68°C)	15 seconds
Beef Roast (medium), Pork Roast, Ham	145°F (63°C)	4 minutes
Poultry, Stuffed Meats, Stuffed Food Products	165°F (74°C)	15 seconds

* See additional cooking times and temperatures in FDA Food Code - paragraph 3-401.11(B)(2)

2001 FDA Food Code

Time and Temperature (continued)

Hot-Holding:

Keeping hot food out of the temperature danger zone

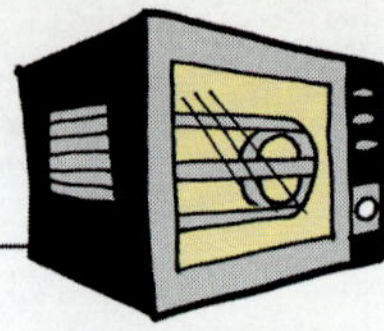

Food Product	Internal Temperature
Hot-holding of all foods	135°F (57°C) or above

Cold Food Holding:

Keeping cold food out of the temperature danger zone

Food Product	Internal Temperature
Cold-holding of all foods	41°F (5°C) or below

Cooling Hot Foods:

Rapid reduction of temperature through and out of the temperature danger zone

Part	Internal Temperature	Times
Hot Food Cooling part 1	From 135° to 70°F (57° to 21°C)	2 hours or less
Hot Food Cooling part 2	From 135° to 41°F (57° to 5°C) or below	Within 6 hours or less

Frozen Food Holding: **Keeping food solidly frozen**

Food Product	Internal Temperature
Frozen Food	Solidly Frozen 0°F (-18°C) recommended

Reheating: **Bringing food back up to serving temperature**

Method	Internal Temperature	Times
Reheating	165°F (74°C) or above	Within 2 hours
Reheat commercially processed, intact packaged, ready-to-eat food	135°F (57°C) or above	Within 2 hours

Remember, there's NEVER been a case of foodborne illness that couldn't have been prevented!

2001 FDA Food Code

Keep Cold Foods Cold and Hot Foods Hot!

Frozen foods should be kept solidly frozen until they are ready to be used. Freezing helps to retain product quality. Proper frozen food temperatures do not permit disease-causing and spoilage microorganisms to grow. Cold temperatures also help to preserve the color and flavor characteristics. Frozen foods can be stored for long periods of time without losing their wholesomeness and quality.

Refrigerated foods are held cold, not frozen. Cold foods should be maintained at 41°F (5°C) or below. Do not forget that some harmful bacteria and many spoilage bacteria can grow at temperatures below 41°F (5°C), although their growth is very slow. By keeping cold foods at 41°F (5°C) or below, you can reduce the growth of most harmful microorganisms and extend the shelf life of the product. For maximum quality and freshness, hold cold foods for the shortest amount of time possible.

Keep cold foods at 41°F (5°C) or below.

Fish displayed on ice to maintain 41°F (5°C) or below

Applying heat is another method used to preserve food. Heat food to proper temperatures to destroy harmful bacteria. Established safe cooking temperatures are based on the type of food and the method used to heat the product. Cooked foods, as well as those foods that have been cooled and then reheated, must be maintained at 135°F (57°C) or above until used. You must keep foods hot to stop growth of harmful bacteria.

There are times during food production when foods must be in the temperature danger zone. Recognize the time spent in the temperature danger zone should be minimal for potentially hazardous items.

Improper holding temperatures is the No. 1 factor that leads to foodborne illness.

Improper holding temperatures is the No. 1 contributing factor that leads to foodborne illness. Spores of certain bacteria like *Clostridium botulinum*, *Clostridium perfringens*, and *Bacillus cereus* can survive cooking temperatures. Remember, if spores survive and are exposed to ideal conditions, they can again become vegetative cells and begin to grow in foods.

The *FDA Food Code* contains cooling guidelines that permit foods to be in the temperature danger zone for a total of six hours. The *FDA Food Code* specifically states foods must be cooled from 135°F (57°C) to 70°F (21°C) in two hours, and from 135°F (57°C) to 41°F (5°C) or less within six hours.

To destroy many of the bacteria that may have grown during the cooling process, reheat foods to 165°F (74°C) within two hours to prevent the number of organisms from reaching levels that can cause foodborne illness.

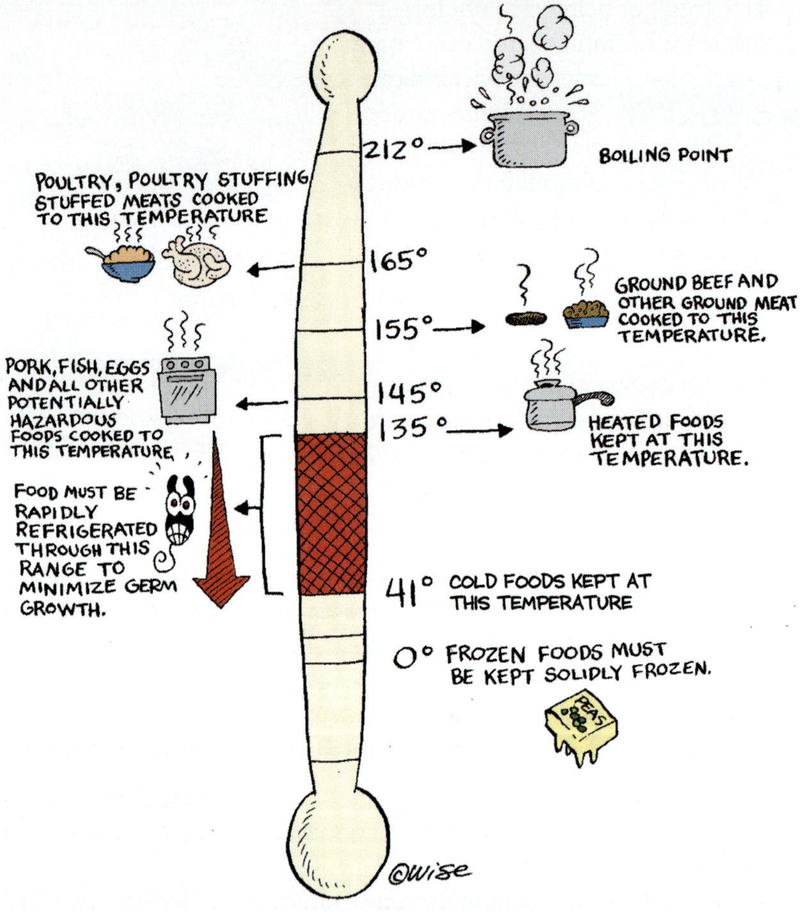

Keep it Hot, Keep it Cold, or Don't Keep It

Ways to Thaw Food

The preferred method for thawing foods is in the refrigerator at 41°F (5°C) or below. This prevents the food from entering the food temperature danger zone.

Other acceptable methods for thawing include using a microwave oven, as a part of the cooking process, or submerging under cool running water [70°F (21°C)] for a controlled amount of time. Proper thawing reduces the chances for bacterial growth, especially on the outer surfaces of food.

More detailed strategies for minimizing the amount of time a food is in the temperature danger zone during cooling, thawing, and food preparation will be presented in the next chapter.

The Importance of Handwashing and Good Personal Hygiene

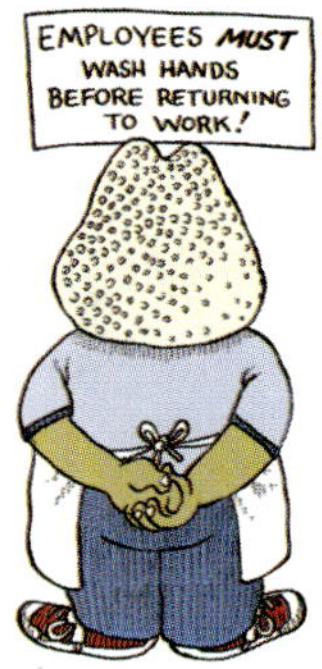

The cleanliness and personal hygiene of food employees are extremely important. If a food employee is not clean, the food can become contaminated. Good personal hygiene is essential for those who handle foods. Desirable behaviors include:

- Knowing when and how to properly wash hands
- Wearing clean clothing
- Maintaining good personal habits (bathing, washing and restraining hair, keeping fingernails short and clean, washing hands after using toilet, etc.)
- Maintaining good health and reporting when sick to avoid spreading possible infections.

Just think of all the things a food handler's hands touch during a typical workday. They may take out the trash, cover a sneeze, scratch an itch, or mop up a spill. When you touch your face or skin, run your fingers through your hair or a beard, use the toilet, or blow your nose, you transfer potentially harmful germs to your hands. *Staphylococcus aureus*, Hepatitis A, and *Shigella spp*. are examples of pathogens that are commonly transferred to foods by bare hand contact.

Personnel involved in food preparation and service must know how and when to wash their hands. Using approved cleaning compounds (soap or detergent), food employees must vigorously rub surfaces of fingers and fingertips, front and back of hands, wrists, and forearms for at least 20 seconds. Remember, soap, warm water, and friction are needed to adequately clean skin. A significant number of germs are removed by friction alone. A brush can be helpful when cleaning hands. However, the brush must be kept clean and sanitary.

When washing hands, thoroughly rinse under clean, warm running water and around fingernails and between fingers. Dry hands using a single-service paper towel, an electric hand dryer, or clean section of continuous rolled cloth towel (if allowed in your area). Do not dry hands on your apron or a dishtowel.

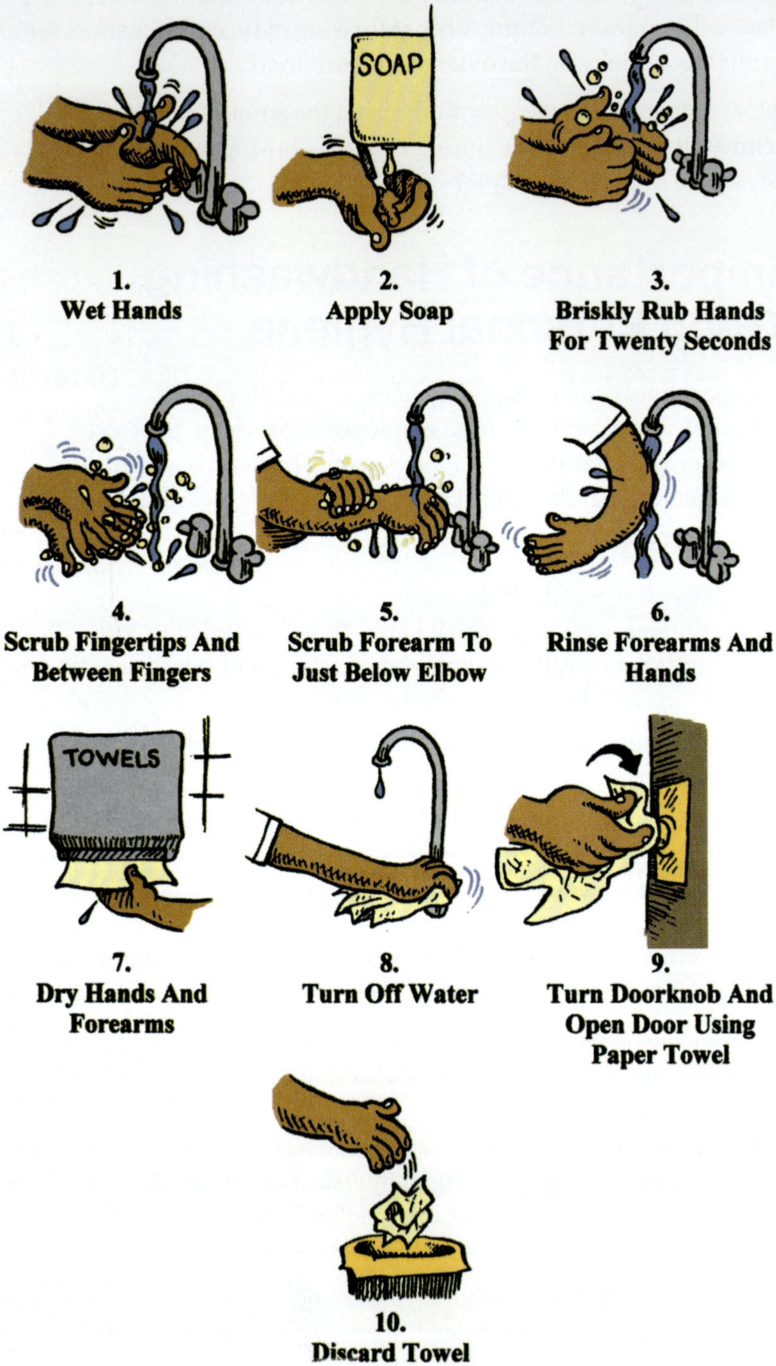

Proper Handwashing Technique

In addition to proper handwashing, fingernails should be trimmed, filed, and maintained so hand washing will effectively remove soil from under and around them. Unless wearing intact gloves in good repair, a food employee must not wear fingernail polish or artificial fingernails when working with exposed food.

According to the *FDA Food Code,* hands shall be washed in a separate sink specified as a handwashing sink. An automatic handwashing facility may be used by food employees to clean their hands. However, the system must be capable of removing the types of soils encountered in the food operation. Food employees may not clean their hands in a sink used for preparation or warewashing, or in a service sink used for the disposal of mop water and liquid waste. Hand sanitizing lotions and chemical hand sanitizing solutions maybe used by food employees in addition to handwashing. Proper washing helps to remove visible hand dirt and the microorganisms it contains. Hand sanitizing lotions must never be used as a replacement for hand washing.

Always Wash Hands:

- Before food preparation
- After touching bare human body parts, except clean hands and clean exposed arms
- After using the toilet
- After coughing, sneezing, using a handkerchief or tissue, using tobacco, eating, or drinking
- During food preparation when switching between raw foods and ready-to-eat products
- After engaging in any activities that may contaminate hands (taking out the garbage, wiping counters or tables, handling cleaning chemicals, picking up dropped items, etc.)
- After caring for or touching service animals or aquatic animals.

It is critical hand sanitizers be formulated with safe and approved ingredients because it is likely a food employee's hands will touch food, food-contact surfaces, or equipment and utensils after using the product.

Handwashing station

Except when washing fruits and vegetables, food employees should not contact exposed ready-to-eat foods with their bare hands. Instead, they should use suitable utensils such as deli tissue, spatulas, and tongs.

Complying with the no bare hands policy is especially important when handling food for sale to highly susceptible populations.

Using Disposable Gloves

Retail food establishments sometimes allow their food employees to use disposable gloves as an extra barrier to help prevent contamination of foods. Gloves can protect food from direct contact by human hands. Gloves must be impermeable, meaning they do not allow anything to penetrate the porous texture of the glove. Use of gloves is often a good idea for foods handled extensively by hands such as deli sandwiches, tacos, or when conducting food demonstrations. You must treat disposable gloves as a second skin. Whatever can contaminate a human hand can also contaminate a disposable glove. Therefore, whenever hands should be washed, a new pair of disposable gloves should be worn. For example, if food handlers are wearing disposable gloves and handling raw food, they must discard those gloves, wash their hands, and put on a fresh pair of gloves before they handle ready-to-eat foods. This procedure is very important in the seafood department and other areas where raw and ready-to-eat foods are handled and displayed.

Food handlers must not handle money with gloved hands unless they immediately remove and discard the gloves. Because money is handled and exchanged by human hands, it is often contaminated with bacteria. Employees must also put on a clean pair of gloves after they complete cleaning, mopping, and similar activities within their department.

If an employee removes gloves by rolling them inside out, the inner surface of the glove is very contaminated from his or her skin. Again, if you take disposable gloves off, throw them away. Never reuse or wash disposable gloves—always throw them away after each use.

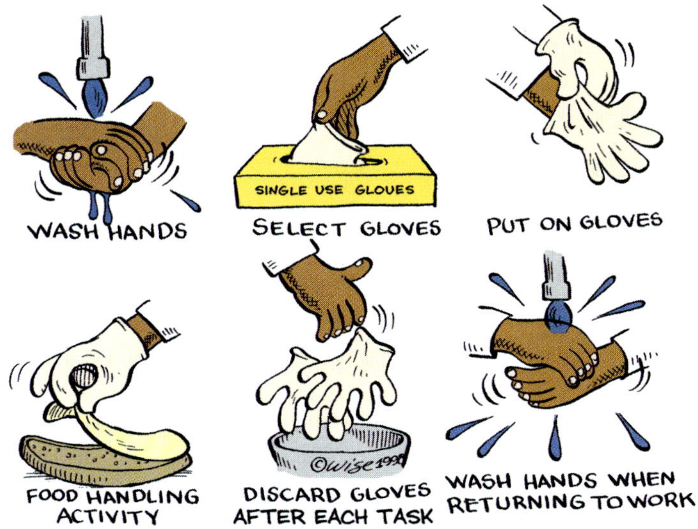

Proper Use of Disposable Gloves

Personal Habits

Personal hygiene means good health habits including bathing, washing hair, wearing clean clothing, and frequent hand washing. A food employee's fingers may be contaminated with saliva during eating and smoking. Saliva, sweat, and other body fluids can be harmful sources of contamination if they get into food.

Supervisors should enforce rules against eating, chewing gum, and smoking in food preparation, service, and warewashing areas. The *FDA Food Code* permits food employees to drink beverages to prevent dehydration. The beverage must be in a covered container. The container must be handled in a way that prevents contamination of the employee's hands, the container, exposed food, equipment, and single-use articles.

Jewelry, including medical information jewelry on hands or arms, has no place in food production and warewashing areas. Rings, bracelets, necklaces, earrings, watches, and other body part ornaments can harbor germs that can cause foodborne illness. Jewelry can also fall into food causing a physical hazard. A plain wedding band that does not contain a stone is the only piece of jewelry that may be worn in food production and warewashing areas.

Outer Clothing and Apparel

Work clothes and other apparel should always be clean. The appearance of a clean uniform is more appealing to your customers.

Things You Can Do to Prevent Food Contamination:

- Wear clean clothing.
- If your clothing is contaminated, change into a new set of work clothes.
- Change your apron between working with raw foods and ready-to eat foods. Aprons should be left in the department when going on break or to the restroom.
- Don't dry or wipe your hands on your apron.
- Wear a hat, hair coverings or nets, and beard restraints to discourage you from touching your hair or beard. These restraints also prevent hair from falling into food or onto food-contact surfaces.
- Keep in mind, however, protective apparel is similar to a disposable glove. They no longer protect food when contaminated.

Personal Health

In an attempt to reduce the risk caused by sick food employees, the *FDA Food Code* requires employees to report to the person in charge when they have been diagnosed with:

- *Salmonella* Typhi
- *Shigella spp.*
- Shiga toxin-producing *Escherichia coli*
- Hepatitis A virus
- Symptoms of intestinal illness or flu-like symptoms (such as vomiting, diarrhea, fever, sore throat, or jaundice) or a wound containing pus such as a boil or infected cut that is open or draining.

If a food employee is directly or indirectly exposed to *Salmonella* Typhi, *Shigella spp.*, Shiga toxin-producing *Escherichia coli*, or Hepatitis A virus, it must be reported to the supervisor. All these agents are easily transferred to foods and are considered severe health hazards. The person in charge will notify the regulatory authority that a food employee is diagnosed with an illness due to *Salmonella* Typhi, *Shigella spp.*, Shiga-toxin producing *Escherichia coli*, or Hepatitis A virus.

If a food employee has been exposed to any of these agents, he or she may be excluded from work or be assigned to restricted activities that do not involve contact with food. Employees diagnosed with one of these diseases must not handle exposed food or have contact with clean equipment, utensils, linens, or unwrapped single-service utensils. Exclusions and restrictions for those exposed to or experiencing symptoms of a foodborne illness are specified in Appendix D of this book.

SICK EMPLOYEES CANNOT WORK WITH FOOD

Staphylococcus aureus bacteria are often found in infected wounds, cuts, and pimples. Infected wounds should be cleaned and completely covered by a dry, tight-fitting, impermeable bandage. Cuts or burns on a food employee's hands must be thoroughly cleaned, bandaged, and covered with a clean disposable glove.

When a foodborne illness outbreak occurs and an employee is suspected of carrying the disease, he or she must not work until certified as safe by a licensed physician, nurse practitioner, or physician assistant. The health and hygiene of food employees are extremely important in safe food management.

To date, there has not been a medically documented case of Acquired Immune Deficiency Syndrome (AIDS) transmitted by food. Therefore, AIDS is not considered a foodborne illness.

The Americans with Disabilities Act (ADA) prohibits discrimination against people with disabilities in jobs and public accommodations. Employers may not fire or transfer individuals who have AIDS or test positive for the HIV virus away from food-handling activities. Employers must also maintain the confidentiality of employees who have AIDS or any other illness.

Cross Contamination

Contaminated food contains germs or harmful substances that can cause foodborne illness. **The transfer of germs from one food item to another is called cross contamination.** This commonly happens when germs from raw food are transferred to a cooked or ready-to-eat food via contaminated hands, equipment, or utensils. For example, bacteria from fish can be transferred to ready-to-eat seafood (such as cooked shrimp and imitation crab) in a display case where the products are not properly separated or by an employee's gloves. Cutting boards are also common vehicles for cross contamination. The leftover juices from raw chicken can lead to contamination of ready-to-eat salad items.

Cross contamination also happens when raw foods are stored above ready-to-eat foods. Juices from the raw product can drip or splash onto a ready-to-eat food. This poses a serious health risk because ready-to-eat items will not be cooked to destroy microorganisms prior to being eaten.

In a retail food establishment, germs can be transferred by a food employee, equipment and utensils, or another food. The following preventative measures can be used to eliminate the possibility of cross contamination between products:

- Always store cooked and ready-to-eat foods over raw products
- Keep raw and ready-to-eat foods separate during storage
- Use good personal hygiene and hand washing
- Keep all food-contact surfaces clean and sanitary
- Avoid bare hand contact with ready-to-eat food
- Keep species of meat and poultry separate
- Use separate equipment, such as cutting boards, for raw foods and ready-to-eat foods (color coding may be helpful for this task)
- Use clean, sanitized equipment and utensils for food production
- Prepare ready-to-eat foods first—then raw foods
- Prepare raw and ready-to-eat foods in separate areas of the establishment.

Use color-coded cutting boards for different types of foods.

Always keep raw foods separate from ready-to-eat foods. In the refrigerator, ready-to-eat foods must be stored above raw foods. Display cases, such as those used to display seafood items, should be designed to keep raw and cooked food items separate. In addition, separate buckets for in-place sanitizing solutions and wiping cloths should be used for cleaning food-contact surfaces in raw and ready-to-eat food production areas.

Other Sources of Contamination

Raw fruits and vegetables should be treated like ready-to-eat foods. Always wash these foods before use. Washing removes soil and other contaminants. Chemicals may be used to wash raw whole fruits and vegetables. These chemicals must be nontoxic and meet the requirements set forth in the Code of Federal Regulations under title 21 CFR 173.315.

Utensils used to dispense and serve foods can also be a source of food contamination. Utensils should be properly labeled to identify the type of food they are used to dispense. During hot- or cold-holding of foods, the utensil should be stored in the food. This helps to prevent contamination from employees or customers in self-service areas. It also keeps the utensil that contains food, out of the temperature danger zone. The dispensing utensils (scoops) for ice and dry bulk foods should be clean and kept in an area protected from contamination. Scoops or tongs used in customer service areas also need to be labeled and kept clean.

Customer self-service areas for bakery items and candy are becoming more popular. Contamination by food allergens, such as peanuts, has occurred in these areas. Be sure the scoop used for peanut butter cups is not used for hard candies. To assure dispensing utensils are used for the intended foods, many retail food establishments will attach the dispensing utensil to the display case with a cable or chain.

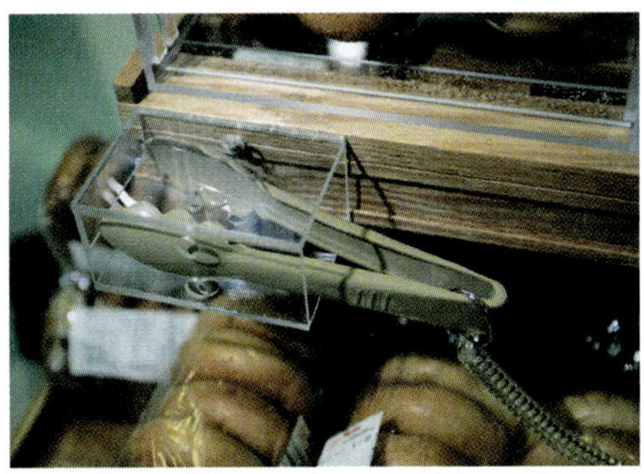

Attach utensils to self-service display cases.

Wash hands or change gloves after touching live animals, such as live lobsters.

Animals are not allowed in retail food establishments unless they are being used for support or special service (i.e., guide dogs for the blind). It is very important food handlers do not touch animals during food preparation and service. If employees should touch an animal for any reason, they must wash their hands before returning to work. For example, if you should remove a live lobster from a tank for sale, you should change your gloves and/or wash your hands properly.

Germs from a employee's mouth can be transferred to food when the employee uses improper tasting techniques. A food employee may not use a utensil more than once to taste food that is to be sold or served.

Animals, rodents, and pests are common sources for food contamination. Rodents and pests usually enter retail food establishments during delivery or when garbage facilities are not properly maintained. A good integrated pest management (IPM) program should be established and maintained in every retail food establishment. You will learn about IPM programs in Chapter 7, *Environmental Sanitation and Maintenance.*

Make Sure the Work Area is Clean and Sanitary

Anything that comes in contact with food must be clean and sanitary. This includes human hands, equipment, utensils, storage and holding areas, and self-service areas for customers. In order to protect food from contamination, effective cleaning and sanitizing procedures must be implemented and monitored. The goal of cleaning is to remove visible soil. The goal of sanitizing is to reduce the number of harmful microbes that may be present on a clean surface. Chapter 6 of this book is dedicated to describing different elements of cleaning and sanitizing programs. While cleaning and sanitizing are important in all areas of a retail food establishment, they are especially critical in areas where ready-to-eat foods are handled and displayed.

Check It Out

Back to the Story... *The case at the beginning of this chapter raises several issues concerning ready-to-eat food preparation areas. The true source of foodborne illness was the use of the same knife and the same cutting board for the raw chicken and the cut-up fruit. The cutting board was turned over, but not washed and sanitized. Though sanitizing solutions were available in the bakery and deli, it was not clear if employees cleaned and sanitized food-contact surfaces on a regular basis.*

- ✓ Temperature abuse can occur during receiving, storing, cooking, cooling, reheating, hot-holding, and cold-holding of foods. Depending on the food type, there are specific temperature requirements to assure food safety. These requirements are discussed in the next chapter. A good rule to follow in any retail food establishment is, "Keep It Clean! Keep It Hot! Keep It Cold! Or Don't Keep It!"
- ✓ Food employees should always use good health and hygiene practices. Clean clothing, hair restraints, and proper handwashing practices are fundamental in safe food management. Effective supervision and enforcement of proper procedures form the foundation of a successful food operation.
- ✓ Control contamination with proper cleaning and sanitizing. Avoid cross contamination from one food to another. Keep foods separate and store raw foods below cooked and ready-to-eat foods. Always use proper food-handling techniques.
- ✓ Keep foods at proper temperature, use good personal hygiene, and control contamination and cross contamination. These are the essentials of safe food management.

Almost All Foodborne Illnesses are Linked to:

- Improper holding temperature
- Poor personal hygiene
- Contaminated equipment
- Inadequate cooking
- Food from an unsafe source.

Quiz 3

Choose the **Best** answer for each question.

1. Packages of luncheon meat kept at temperatures above 50°F (10°C) for more than six hours are:
    a. Safe to eat.
    b. Cross contaminated.
    c. Time and temperature abused.
    d. Adulterated.
2. Good personal hygiene includes:
    a. Using hand sanitizers instead of handwashing.
    b. Keeping hands and clothes clean and sanitary.
    c. Wearing attractive uniforms.
    d. Cleaning and sanitizing food-contact surfaces.
3. Before changing to ready-to-eat foods, a cutting board used to cut up raw chicken must be:
    a. Cleaned and sanitized.
    b. Turned over to use the other side.
    c. Dried thoroughly with a paper towel.
    d. Rinsed in cool running water.
4. After proper cooking, all foods that are to be held hot must be held at:
    a. 165°F (74°C) or above.
    b. 135°F (57°C) or above.
    c. Room temperature until served.
    d. 120°F (49°C) or above.
5. After proper cooking, all foods that are to be held cold must be:
    a. Cooled quickly and held at 41°F (5°C) or below.
    b. Cooled quickly and held at 70°F (21°C) or above.
    c. Stored at room temperature until served.
    d. Cooled slowly and held at 50°F (10°C) or below.

Anwers are available in Appendix A.

References/Suggested Readings

Centers for Disease Control and Prevention (2000). *Surveillance for Foodborne Disease Outbreaks—United States, 1993-1997.* March 17, 2000. Atlanta, GA; U.S. Department of Health and Human Services.

Code of Federal Regulations. 2001. *21 CFR 173.315 Secondary Direct Food Additives Permitted in Food for Human Consumption - Chemicals used in Washing or to Assist in the Peeling of Fruits and Vegetables.* U.S. Government Printing Office. Washington, D.C.

Council for Agricultural Sciences and Technology (1995). Prevention of Food-borne Illness. Dairy. *Food and Environmental Sanitation.* Vol. 15(6), 341-367.

Food and Drug Administration (2001). *2001 Food Code.* U.S. Public Health Service. Washington, DC.

Food Safety and Inspection Service (1996). *Nationwide Broiler Chicken Microbiologic Baseline Data Collection Program, 1994-1995.* U.S. Department of Agriculture. Washington, DC.

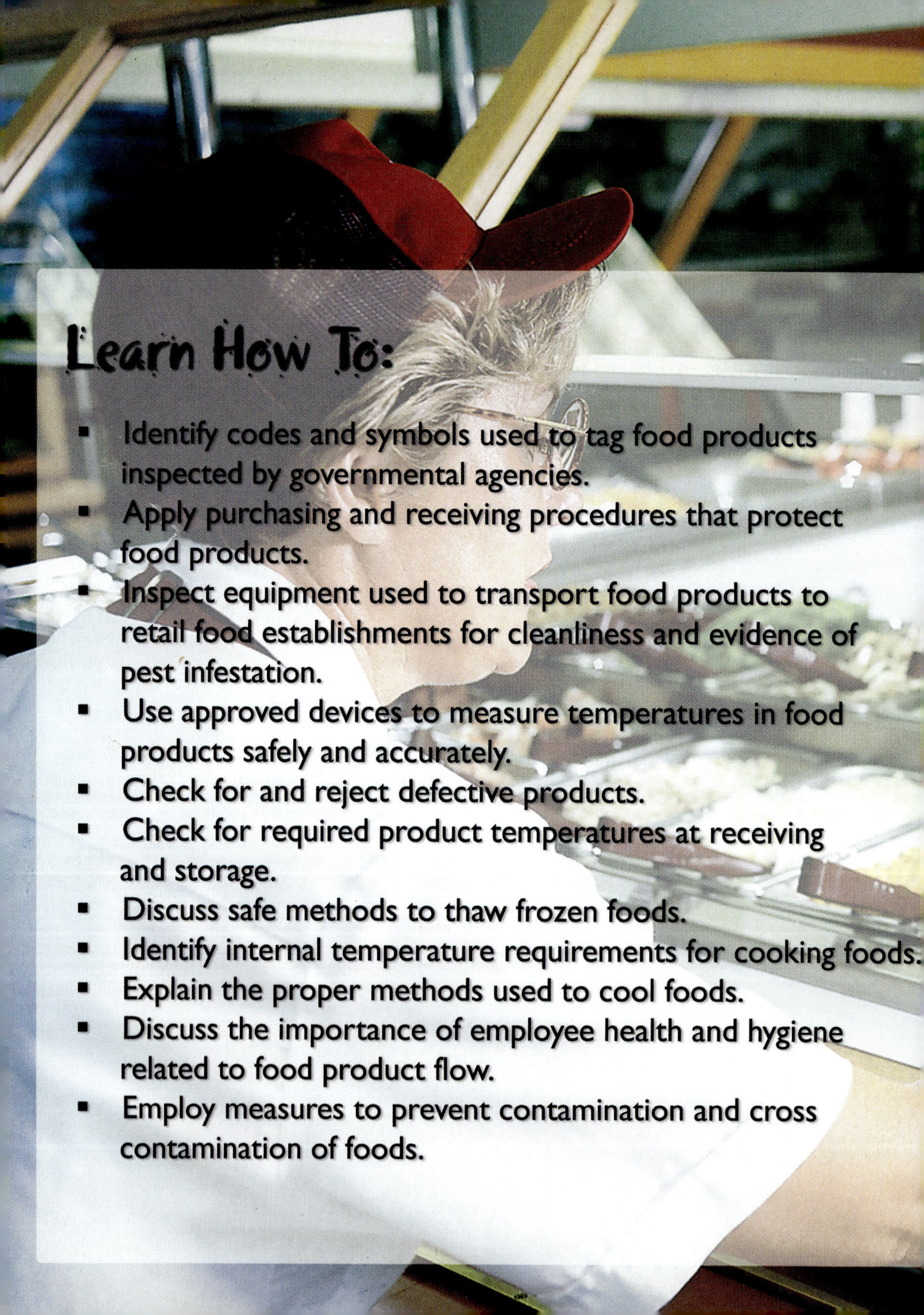

Learn How To:

- Identify codes and symbols used to tag food products inspected by governmental agencies.
- Apply purchasing and receiving procedures that protect food products.
- Inspect equipment used to transport food products to retail food establishments for cleanliness and evidence of pest infestation.
- Use approved devices to measure temperatures in food products safely and accurately.
- Check for and reject defective products.
- Check for required product temperatures at receiving and storage.
- Discuss safe methods to thaw frozen foods.
- Identify internal temperature requirements for cooking foods.
- Explain the proper methods used to cool foods.
- Discuss the importance of employee health and hygiene related to food product flow.
- Employ measures to prevent contamination and cross contamination of foods.

CHAPTER 4

Following the Food Product Flow

Improperly held bakery items...

An employee in the bakery department did not follow proper food-labeling procedures for pastry products. He was supposed to label each tray of pastries using the first-in, first-out method. As a result of his error, a group of egg custard-filled éclairs was left on the finishing rack for over six hours at room temperature before another employee at a shift change noticed it. She quickly recognized the need for refrigeration and immediately set the unlabeled éclairs in the cold case. However, she did not know how long the items had been kept in the rack at room temperature. What is the problem here and what would have been the best solution?

Buying from Approved Sources

Food quality and safety begin with foods and ingredients from approved sources—processors and suppliers that comply with federal, state, and local food safety laws and regulations. These sources are routinely inspected to make sure they follow good manufacturing practices. Foods prepared in a private home are not from an approved source and must not be used or sold in retail food establishments. Use of "home-canned" food is prohibited because of the high risk of foodborne illness, especially botulism.

Inspecting Delivery Vehicles

Suppliers must deliver food products to your retail food establishment in vehicles that are clean and in good repair. Delivery vehicles must also:

- Maintain perishable and potentially hazardous foods at safe temperatures during transport
- Be loaded in a manner that separates food items from non-food items (detergents, household cleaners, and pesticides) to prevent contamination and cross contamination
- Protect food packages from becoming damaged and torn during transit.

When delivery vehicles arrive at your establishment, receiving personnel should inspect them. Specific items to look for during these inspections are:

- Cleanliness of the cargo area
- Temperature of refrigerated and frozen storage areas (if applicable)
- Proper separation of food and non-food items
- Signs of insect, rodent, or bird infestations
- Damaged packages that might result in contamination of food items.

Determining Food Quality

The senses of smell, touch, sight, and, sometimes, taste are frequently used to evaluate the quality of food received. As a first step, foods should be observed for color, texture, and visual evidence of spoilage. Quite often, spoilage is easily seen as slime formation, mold growth, and discoloration.

Inspect foods when they are received.

Spoiled foods frequently give off foul odors indicative of compounds such as ammonia and hydrogen sulfide (the smell of rotten eggs). These odors, caused by the breakdown of proteins through bacterial action, are usually very easy to smell.

Spoilage due to yeasts produces bubbles and an alcoholic flavor or smell. Milk develops an acidic taste and is often bitter or rancid when it spoils.

The quality and safety of a food are affected by many factors. A food that shows no signs of spoilage may not always be safe. Spoilage cannot be used as the only indicator of food safety.

Measuring Temperatures at Receiving and Storage

Temperature-measuring devices are used in retail food establishments to measure temperatures of food, water, and the air of food storage areas (refrigerators, ovens, etc.).

Maintaining safe product temperature is a critical part of your food safety system.

Cold- or hot-holding equipment used for storing potentially hazardous foods should be equipped with an indicating or recording thermometer to measure the temperature of the storage environment. Equipment thermometers are either built into a piece of equipment or are fastened onto shelving or other apparatus. Set them where they can be easily read. Place the sensor portion of the thermometer in the warmest part of a refrigeration unit or in the coolest part of a hot food storage unit. Some older equipment may need modification if they were designed with the sensor located in the discharge air.

Following the Flow of Food

The flow of food at a retail food establishment begins with receiving and storage. As foods are delivered, they are inspected by receiving personnel and then placed quickly into storage.

From storage, foods and ingredients flow into the preparation and handling stages of production or directly into a display case. Some common preparation steps in retail food establishments include grinding, cutting, trimming, dispensing and packaging. A simple diagram showing the overall flow of food in a retail food establishment is presented below.

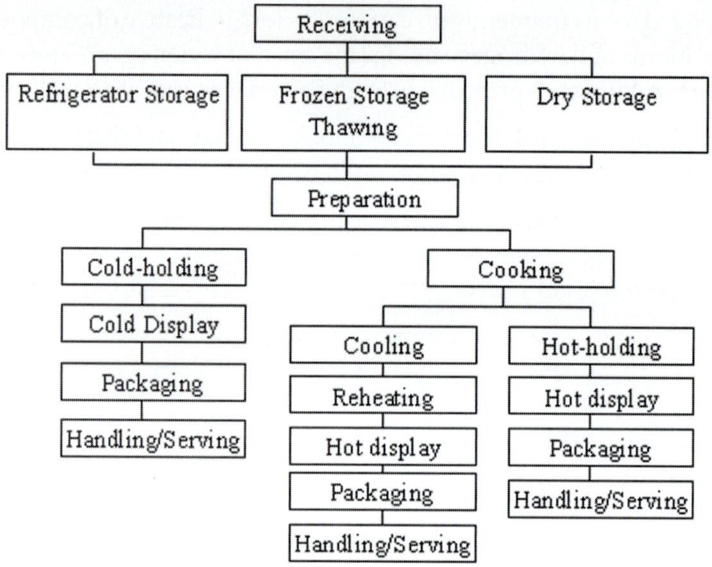

Food flow diagram

Receiving

Inspect all incoming food supplies to make sure they are at the proper temperature, in sound condition, and free from filth or spoilage. Always check product containers for tears, punctures, dents, or other signs of damage. Poor receiving procedures increase the chance of:

- Theft
- Acceptance of underweight merchandise
- Contamination

- Waste
- Acceptance of products that do not meet specifications.

Schedule deliveries for off-peak times and have enough staff and space on hand to receive products quickly and correctly. Move incoming shipments to storage as soon as they arrive. Merchandise that is damaged, spoiled, or otherwise unfit for sale or use must be properly disposed of, held by the store for credit, or returned to the distributor. Distressed merchandise must be put aside to prevent contamination of other foods, equipment, utensils, linens, or single-service or single-use articles.

Merchandise that is damaged, spoiled, or otherwise unfit for sale should be rejected by receiving clerks. Store rejected foods away from good food.

Whatever the size of the establishment, receiving requires:

- Prompt handling
- Quality control procedures
- Trained staff who know:
    - ▲ Product specifications
    - ▲ Coding
    - ▲ Accurate checks on product temperatures
    - ▲ Proper handling of rejected merchandise.

Packaged Foods

The foods in a retail food establishment come in many different types of packages. Examples of food packaging include cans, bottles, jars, pouches, tubs, trays, bags and boxes. The common purpose of the package is to:

- Protect the contents from contamination
- Provide a source of information about its nutritional contents
- Provide advertising material

- Make the product more convenient for customers to transport, prepare, and serve.

Dry foods such as flour, sugar, rice, and beans are commonly packaged in bags. These are not potentially hazardous foods and can be safely stored at room temperature. At receiving, packages of dry food must be inspected for tears, signs of damage and contamination by chemicals and other substances, and infestations that could cause foodborne illness or product loss.

Reduced Oxygen Packaging (ROP)

For many foods, oxygen in the air can increase chemical breakdown and microbial spoilage. Many food processors use **reduced oxygen packaging** to help overcome the effects of oxygen and preserved foods. Some examples of reduced oxygen packaging include vacuum packaging, modified atmosphere packaging, and sous vide foods.

Vacuum packaging removes air from a package and hermetically seals the package so a near-perfect vacuum remains inside. The term hermetic refers to a container completely heat-sealed to prevent the entry and loss of gases and vapors. The most commonly used hermetic packages are metal cans and glass jars. These containers will also stop the entry of bacteria, yeasts, molds and other types of contamination as long as they remain undamaged.

Upon receiving, metal cans must be checked for defects.

Defective Cans

Leaking or Bulging

- Do not accept cans if they leak or bulge at either end
- Swollen ends on a can indicate gas is being produced inside
- Gas may be caused by a chemical reaction between the food and the metal in the container or by the bacteria and other microbes inside the can.

(cont.)

Defective Cans (Continued)

Dented

- Dents in cans do not harm the contents unless they have actually penetrated the can or the seam
- Dents found in the side seams or end seams of a can are the most important
- Do not accept cans if damage to these areas can affect the physical integrity of the can and may allow microorganisms to enter through tiny pinhole leaks
- Shipments with many dented cans or torn labels indicate poor handling and storage procedures by the supplier.

Rusty

- Rust does not harm contents unless it has penetrated the can or seam
- Rusty cans indicate exposure to excess moisture.

Modified atmosphere packaging (MAP) helps preserve foods by replacing some or all of the oxygen inside the package with other gases such as carbon dioxide or nitrogen. MAP is used with a wide range of products including meat, fish, pre-cut lettuce, baked products, cheese, coffee, nuts, and dried fruit. The MAP process has been successful in extending the shelf life of many products, and it reduces the amount of additives and preservatives required to prevent deterioration of the food. However, since low oxygen environments can create conditions conducive to the growth of anaerobic bacteria like *Clostridium botulinum* (see Chapter 2 on page 39), proper handling of MAP foods is essential. MAP technology requires adequate refrigeration be maintained during the entire shelf life of potentially hazardous foods. During the receiving process, employees must

Example of MAP food

inspect potentially hazardous foods in modified atmosphere packages to make sure the package is in sound condition and the food is at 41°F (5°C) or below when it arrives at the establishment.

Sous vide is a French term for "without air." It is a specialized packaging process where fresh raw foods are sealed in plastic pouches and the air is removed by vacuum. The pouches are then cooked at a low temperature and rapidly cooled to 38°F (3°C) or below or frozen. The low cooking temperature of the sous vide process kills spoilage microorganisms. However, it does not destroy *C. botulinum* bacteria and their spores. Therefore, potentially hazardous foods processed using sous vide technology must be kept out of the food temperature danger zone during transport, storage, and display. Receiving personnel must inspect sous vide products to make sure the package is in sound condition and the food is at 41°F (5°C) or below as it arrives at the establishment. Some sous vide foods may require even colder temperatures during receipt and cold storage. Check with the manufacturer about cold-holding temperatures for any sous vide foods you purchase. If extended shelf life is desired, a temperature of 38°F (3°C) or lower must be maintained at all times. After receiving, the products should be moved quickly into refrigerated storage until ready for use.

Common ROP Packaging Choices

- **Cook-chill** —a process that uses a plastic bag filled with hot cooked food from which air has been forced out and which is closed with a plastic or metal crimp.
- **Controlled Atmosphere Packaging (CAP)** —a system that maintains the desired atmosphere within a package throughout the shelf life of a product by the use of agents to bind or scavenge oxygen or a small packet containing compounds to emit a gas.
- **Modified Atmosphere Packaging (MAP)**—a process that employs a gas flushing and sealing process or reduction of oxygen through respiration of vegetables or microbial action.
- **Sous Vide** —a process where fresh raw foods are sealed in a plastic pouch and the air is removed by vacuum. The pouch is cooked at a low temperature and rapidly cooled to 38°F (3°C) or below or frozen.
- **Vacuum Packaging** —reduces the amount of air from a package and hermetically seals the package so a near-perfect vacuum remains inside.

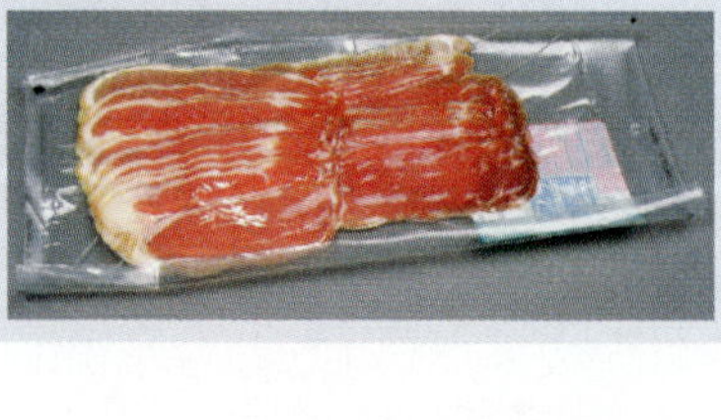

Benefits to ROP Packaging

- It creates a largely oxygen-free environment that prevents the growth of aerobic bacteria, yeast, and molds largely responsible for the off odors, slime, texture changes, and other forms of spoilage
- It prevents chemical reactions that can produce off odors and color changes in foods
- It reduces product shrinkage by preventing water loss.

Some products cannot be packed by ROP unless the retail food establishment is approved for the activity and inspected by the appropriate regulatory authority. These products include raw and smoked fish, soft cheeses (ricotta, cottage cheese, and cheese spreads), and combinations of cheese and other ingredients such as vegetables, meat, or fish. Contact your local regulatory authority to obtain a complete set of guidelines for ROP foods and to seek a variance for all food manufacturing/processing operations based on the prior approval of a HACCP plan.

The *FDA Food Code* Requirements For Retail Food Establishments That Use ROP Technology

- **Have a HACCP plan** —The retail food establishment must have a HACCP plan in place for the ROP operation that details how the seven principles (see Chapter 10) are incorporated into the operation's food safety management system.

- **Use two microbial growth barriers** —Because *Clostridium botulinum* is a significant health hazard, the establishment is required to have two microbial growth barriers. Time/temperature are commonly used as one barrier, but they need to be coupled with the use of pH, A_W, or product formulation to assure *Clostridium botulinum* will not grow inside the package.

- **Maintain foods at proper temperature** —All potentially hazardous foods in ROP that rely on refrigeration as a barrier to microbial growth must be maintained at 41°F (5°C) or below.

- **Set shelf life** —The refrigerated shelf life is to be no more than 14 days from packaging to consumption or the original manufacturer's "sell by" or "use by" date, whichever comes first.

(cont.)

The *FDA Food Code* Requirements For Retail Food Establishments That Use ROP Technology (continued)

- **Proper label warnings** —Packages must be prominently and conspicuously labeled on the principal display panel with the instructions to maintain the food at 41°F (5°C) or below and to discard refrigerated food if within 14 days of its packaging it is not served for on-premises consumption or consumed if served or sold for off-premises consumption.

- **Use-by or sell-by dates** —Each container must bear a use-by or sell-by date. This date cannot exceed 14 days from packaging or repackaging to consumption, except the time the product is maintained frozen or the original manufacturer sell-by or use-by date whichever occurs first.

- **Employee training** —Employees responsible for the ROP process must receive training that will enable them to understand the key components of the process, the equipment used, and the specific procedures that must be followed to ensure critical limits identified in the HACCP plan have been met. The training program must also outline the employee's responsibilities for monitoring and documenting the process and describe what corrective actions they must take when critical limits are not met.

Food **irradiation** is a preservation technique used by some food processing industries. This process involves exposing food to certain forms of radiation in order to destroy disease-causing microorganisms and delay spoilage.

Radura symbol

The FDA has approved food irradiation for a variety of foods including fruits, vegetables, grains, spices, poultry, pork, lamb, and, more recently, ground beef.

The acceptance of irradiated foods has been limited due to customers' concerns about the safety of foods preserved in this manner. Contrary to many myths, irradiated food is not radioactive and does not pose a risk to the health and safety of people who eat it. Foods processed with irradiation are just as nutritious and flavorful as other foods that have been cooked, canned, or frozen.

Federal law requires irradiated food to be labeled with the international symbol for irradiation called a "radura." This symbol must be accompanied by the words "Treated with Irradiation" or "Treated with Radiation."

Irradiation of food can effectively reduce or eliminate pathogens and spoilage microbes while maintaining the quality of most foods. This is a technology proven safe and should be welcomed by customers as an effective food preservation technique.

Red Meat Products

Raw meat

Most red meat and meat products sold in the United States come from cattle (beef), calves (veal), hogs (ham, pork, and bacon), sheep (mutton), and young sheep (lamb). These products are inspected for wholesomeness by officials of the U.S. Department of Agriculture (USDA) or state agencies. Animals must be inspected for wholesomeness to make certain they are free of disease and unacceptable defects. The USDA also offers voluntary meat grading services. Grades for meat represent the culinary quality or palatability of the meat and are not measures of product safety. The figure below contains examples of inspections and grading stamps applied to products approved for use.

Inspection **Grade**

USDA inspection and grade stamps for beef, veal, and lamb

Meat and meat products sold in retail food establishments are available in several forms such as fresh, frozen, cured, smoked, dried, and canned. Since raw meats are potentially hazardous foods, never accept them if there is any sign of contamination, temperature abuse, or spoilage.

Reject fresh meat if the product temperature exceeds 41°F (5°C) at delivery.

- Reject fresh meat if the product temperature exceeds 41°F (5°C) at delivery. Fresh meat should be firm and elastic to the touch and have characteristic aromas. Off odors are frequently indicators of spoilage. Sliminess is another characteristic of spoilage and is caused by bacterial growth on the surface of meat. Control of factors that cause spoilage and sliminess also extends the shelf life of meat products and reduces shrink loss.

- Frozen meats should be solidly frozen when they arrive at your store. Look for signs of freezing and thawing and refreezing such as frozen blood juices in the bottom of the container or the presence of large ice crystals on the surface of the product. Frozen meats should be packaged to prevent freezer burn.

Move fresh meat into refrigerated storage as quickly as possible. Frozen products should be moved from the delivery truck to the freezer while they are still solidly frozen.

Poultry

USDA or state inspectors must inspect all poultry products to make certain they are wholesome and not adulterated. Adulterated food contains filth or is otherwise decomposed and unfit for human consumption. Inspected poultry products carry a USDA seal on the individual package or on bulk cartons.

Inspection   Grade

USDA inspection and grade stamps for poultry

Usually poultry is graded also for quality. Grade A poultry must have good overall shape and appearance, be meaty, be practically free from defects, and have a well-developed layer of fat in the skin.

Poultry products support the growth of disease-causing and spoilage microorganisms. The intestinal tract and skin of poultry may contain a variety of foodborne disease bacteria, including *Salmonella spp.* and *Campylobacter jejuni.* The near neutral pH, high moisture, and high protein content of poultry make it an ideal material for bacteria to grow in and on.

Poultry products are also vulnerable to spoilage caused by enzymes and spoilage bacteria.

Spoilage is indicated by meat tissue that:

- Is soft
- Is slimy
- Has an objectionable odor
- Has stickiness under the wings
- Has discolored or darkened wing tips.

Poultry should be packaged on a bed of ice that drains away from the meat as it melts and held at or below 41°F (5°C).

Spoiled poultry

Game Animals

Game animals are not permitted for sale in retail food establishments unless they meet federal code regulations. This ban does not apply to commercially raised game animals approved by regulatory agencies, field-dressed game allowed by state codes, or exotic species of animals that must meet the same standards as those of other game animals.

Game animals commercially raised for food must be raised, slaughtered, and processed according to standards used for meat and poultry. Common examples of animals raised away from the wild and used for food are farm-raised buffalo, ostrich, and alligator. The USDA inspects the slaughter and processing of this meat in the usual manner.

Eggs

Most retail food establishments sell and use eggs in one form or another. Eggs are usually purchased by federal grades, the most common being AA, A, and B. Grades for eggs are based on exterior and interior conditions of the egg.

Fresh Eggs

The USDA reports approximately 50 billion eggs are sold in the United States each year. *Salmonella enteritidis* bacteria are present in about 1% (approximately 500 million or one in 20,000) of the eggs sold. The bacteria enter the yolk of the egg as it is formed inside the hen. The eggshell surface may contain *Salmonella spp.* bacteria, especially if the shell is soiled with chicken droppings. Even if the shell is not cracked, bacteria can enter through the pores in the egg's shell.

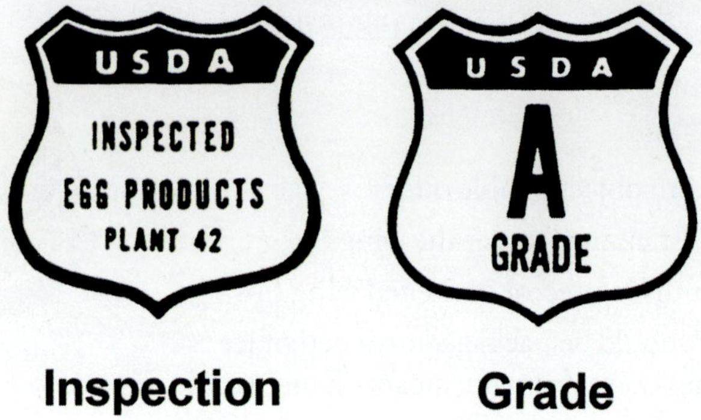

USDA inspection and grade stamps for eggs

Raw shell eggs should be clean, fresh, free of cracks or checks, and refrigerated at an ambient air temperature of 45°F (7°C) or below when delivered. Shell eggs that have not been treated to destroy all viable *Salmonella* shall be stored and displayed in refrigerated equipment that maintains an ambient temperature of 45°F (7°C) or less. These eggs must be labeled to include safe handling instructions. The egg, when opened, should have no noticeable odor, a firm yolk, and the white should cling to the yolk. Reject eggs that are dirty or cracked, and remember washing eggs only increases the possibility of contamination.

Ambient Air - Temperature of surrounding environment

An egg product is defined as an egg without its shell. As a safeguard against *Salmonella spp.*, the FDA requires all egg products, such as liquid, frozen, and dry eggs, be pasteurized to render them *Salmonella*-free. Pasteurized egg products should be in a sealed container and kept at 41°F (5°C). Egg containers should carry labels that verify the contents have been pasteurized. These products are well suited for facilities that offer food to people in highly susceptible populations.

Fluid Milk and Milk Products

This food group includes milk, cheese, butter, ice cream, and other types of milk products. When receiving milk and milk products, make certain they have been pasteurized. Pasteurization destroys all disease-causing microorganisms in the milk and reduces the total number of bacteria, thus increasing shelf life. All market milk must be Grade A quality. Pasteurization also destroys natural milk enzymes that might shorten the shelf life of the products.

Milk that is marked "UHT" pasteurized has been heated to ultra-high temperatures and placed in aseptic packaging. UHT products can be stored safely for several weeks if kept under refrigeration. These products can be stored without refrigeration for short periods of time (Longrèe and Armbruster). Individual creamers are sometimes processed in this manner.

Fluid Milk

Under the Pasteurized Milk Ordinance, fluid milk can be received at 45°F (7°C) or less. It should be refrigerated immediately upon delivery and held at 41°F (5°C) or below. Individual containers of milk should be clearly marked with an expiration date and the name of the dairy plant that produced it. Check the expiration date of all dairy products before using them.

Cheese

Cheese should be received at 41°F (5°C) and checked for the proper color, flavor, and characteristics. Reject the product if it contains mold that is not a normal part of the cheese or if the rind or package is damaged.

Butter

Butter is made from pasteurized cream. Since disease-causing and spoilage bacteria and mold may grow in butter, handle the product as a perishable item. The most common type of deterioration in butter is the development of a strong rancid odor and flavor. Ensure butter has a firm texture, even color, and is free of mold. Packaged butter should be received at 41°F (5°C), intact, and provide protection for the contents.

Fish

Fish includes finfish harvested from saltwater and freshwater, and seafood that comes mainly from saltwater. Seafood consists of molluscan shellfish and crustaceans. Molluscan shellfish include oysters, clams, mussels, and scallops. Crustaceans include shrimp, lobster, and crab. Oysters, shrimp, catfish, salmon, and a few other types of finfish and seafood are being raised on fish farms using a technique called aquaculture.

Fish should be received at 41°F (5°C) or below and shellfish may be received at 45°F (7°C) or below. For better quality and shelf life, the optimum fish and shellfish receiving temperature often ranges from 30°F (-1°C) to 34°F (1.1°C). These products are generally more perishable than red meats, even when stored in a refrigerator or freezer. They are commonly packed in self-draining ice to prevent drying and to maximize the shelf life of the food. The slime covering the outside of fish contains a variety of bacteria that makes them highly susceptible to contamination and microbial spoilage. Fish are also rich in unsaturated fatty acids that are susceptible to oxidation and the development of off flavors and rancidity.

The quality of fish and seafood is measured by smell and appearance. Fresh finfish should have a mild, pleasant odor and bright, shiny skin with the scales tightly attached. Fish with the head intact should have clear, bulging eyes and bright red, moist gills. The flesh of fresh fish should be firm and elastic to the touch.

Fish must be commercially and legally caught or harvested— except when caught recreationally— and approved for sale by the regulatory authority. All fish suppliers and warehouse operations must comply with the seafood HACCP program as required in 21 CFR 123. Ready-to-eat raw, marinated, or partially cooked fish other than molluscan shellfish must be frozen to time and temperature guidelines that meet *FDA Food Code* specifications in order to kill parasites. Records must be retained that show how the product was handled.

Purchase molluscan shellfish from approved sources.

Shellfish must be purchased from sources approved by the Food and Drug Administration and the health departments of states located along the coastline where the shellfish is harvested. Shellfish transported from one state to another must come from sources listed in the Interstate Certified Shellfish Shippers List. The reason for requiring tight control over molluscan shellfish is to reduce the risk of infectious Hepatitis and other foodborne illnesses that may result from eating raw or insufficiently cooked forms of this product. Molluscan shellfish caught recreationally may not be used or sold in retail food establishments.

When received at a retail food establishment, molluscan shellfish should be reasonably free of mud, dead shellfish, and shellfish with broken shells. Damaged shellfish must be discarded.

Molluscan shellfish must be purchased in containers that bear legible source-identification tags or labels fastened to the container by the harvester and each dealer that shucks, ships, or reships the shellstock. Molluscan shellfish tags must contain the following information:

- The harvester's identification number

- The date of harvesting
- An identification of the harvest location or aquaculture site including an abbreviation of the state or country in which the shellfish are harvested
- The shellfish type and quantity
- A statement in bold, capitalized type that says, "**THIS TAG IS REQUIRED TO BE ATTACHED UNTIL CONTAINER IS EMPTY AND THEREAFTER KEPT ON FILE FOR 90 DAYS.**"

Example of shellfish tag

If seafood is suspected of being the source of foodborne illness, the investigating team can use the tags to determine where and when the product was harvested and processed.

For display purposes, shellstock may be removed from the tagged or labeled container in which they are received. The shellstock must be placed on drained ice or held in a display container. Care must be taken to protect shellstock from contamination during display.

Shellstock containers must be kept until every piece of product from that batch is gone.

The identity of the source of shellstock that has been removed from a tagged or labeled container for display must be preserved. This can be accomplished by using an approved record keeping system that keeps shellstock tags or labels in sequence based upon the date when, or dates during which, the shellstock are sold or served. Shellstock from one tagged or labeled container must not be commingled with shellstock from another container before being ordered by the customer. **Commingling** is combining shellfish harvested on different days or from different growing areas as identified on the tag or label, or combining shellfish from containers with different container codes or different shucking dates.

Fruits and Vegetables

Most fruits and vegetables have a short shelf life. They continue to ripen even after they are picked. Therefore, they may become too ripe if not properly handled. Microorganisms found in water and soil can also cause fruits and vegetables to spoil. Fruits and vegetables hold their top quality for only a few days.

Purchase raw fruits and vegetables from approved sources and wash them thoroughly to remove soil and other contaminants before they are cut, combined with other ingredients, cooked, served, or offered for human consumption in a ready-to-eat form.

Whole raw fruits and vegetables that will be washed by customers before they are eaten do not need to be washed at the establishment before they are sold.

Most fresh fruits and vegetables are usually not considered potentially hazardous foods (PHFs). Often the acidity and/or outer skin will prevent the entry and growth of harmful bacteria. However, sprouts and cut melons are now considered PHFs and should be handled in the same manner as other PHFs. Even so, the number of cases of foodborne illnesses linked to these kinds of products has increased in recent years. This is largely due to increased consumption of fresh fruits and vegetables and the emergence of microbes that can cause disease with a low number of organisms. Shiga toxin-producing *Escherichia coli*, *Shigella spp.*, Hepatitis A, and *Cyclospora spp.* can be infective with only a few cells. Therefore, they do not require a potentially hazardous food to multiply. Though not required, whole raw fruits and vegetables may be washed using cleaners. In some cases, an additional anti-microbial rinse (water or bath) is used to reduce the number of microorganisms present on the surface. When these types of chemicals are used, they must meet the requirements in the Code of Federal Regulations (21 CFR 173.315). The fruits and vegetables should also be rinsed to remove as much of the residues of these chemicals as possible.

Some products, like wild mushrooms, may only be used if they have been inspected and approved by a mushroom-identification expert who is approved by the regulatory authority. Beware of fresh mushrooms packaged in Styrofoam trays and covered with plastic shrink-wrap. Mushrooms use up the oxygen inside the package quickly. Unless holes are poked in the plastic wrap that covers the package to permit oxygen inside, oxygen-free conditions may occur that are favorable for the growth of *C. botulinum* bacteria.

Juice

Juice includes the liquid extracted from one or more fruits or vegetables, purees of the edible portions of one or more fruits or vegetables, or any concentrates of such liquid or puree. According to the *FDA Food Code*, this group of foods includes juice as a whole beverage, an ingredient of a beverage, and a puree as an ingredient of a beverage.

Most of the juices sold in retail food establishments are obtained from processors in a prepackaged form. These juices must be obtained from a processor that has a HACCP system in place. In most instances, the processor will pasteurize or otherwise treat the juice to attain 99.999% reduction of the most resistant disease-causing microorganisms of public health significance.

Juice packaged in a retail food establishment must:

- Be treated under a HACCP plan as specified in 8-201.12(B)-(E) of the 2001 *FDA Food Code* to attain a 99.999% reduction of the most resistant microorganisms of public health significance, or if the juice is not treated to destroy pathogens,
- Bear a warning label that informs customers "This has not been pasteurized and, therefore, may contain harmful bacteria that can cause serious illness in children, the elderly, and persons with weakened immune systems."

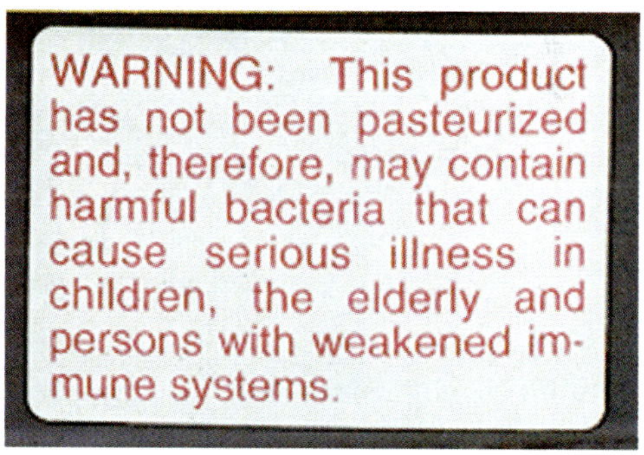

Juice warning label

Frozen Foods

Frozen products must be solidly frozen when delivered. Check the temperature of frozen foods by placing the sensing portion of a thermometer between two packages. Receiving personnel should also look for signs the product has been thawed and refrozen.

Common signs of thawing and refreezing are:

- Large ice crystals or frost on the surface of the food
- Frozen liquid or juice at the bottom of the package
- Mushy soft products.

Reject frozen foods that are not solidly frozen or show signs of temperature abuse.

Storage of Food

Employees must check incoming shipments carefully and quickly move received items to proper storage. Stock rotation is a very important part of effective food storage. A first-in, first-out (FIFO) method of stock rotation helps ensure older foods are used first. Product containers should be marked with a date or other readily identifiable code to help employees know which product has been in storage longest. When expecting food shipments, always make certain the older stock is moved to the front of the storage area to make room for the newly arriving products.

Proper stock rotation

Types of Storage

The three most common types of food storage areas in retail food establishments are the:

- Refrigerator
- Freezer
- Dry storage.

Refrigerated storage is used to hold potentially hazardous and perishable foods. It slows down microbial growth and controls quality by holding foods at 41°F (5°C) or below. Some common types of refrigerated storage equipment found in retail food establishments are:

- Walk-in
- Reach-in
- Coffin
- Upright
- Under-the-counter refrigerators
- Cold display units, among others.

In order to maintain the temperature of potentially hazardous foods at 41°F (5°C) or below, equipment should maintain the air temperature in the storage compartment at about 38°F (3° C). Fish and shellfish are especially vulnerable to spoilage and should be stored at colder temperatures ranging from 30°F (-1°C) to 34°F (1.1°C). Some fruits and vegetables, such as bananas and potatoes, undergo undesirable chemical changes when they are refrigerated. Therefore, although fruits and vegetables are perishable products, not all types should be refrigerated. For perishability, fresh fruits and vegetables requiring refrigeration should be stored at temperatures between 33°F (1°C) and 41°F (5°C). Unpasteurized juices and potentially hazardous fruits and vegetables, like cut melons and sprouts, must be refrigerated at 41°F (5°C) or below.

The unit should maintain air temperature of 38°F (3°C). Check temperatures daily in the warmest part of the unit to ensure accuracy using an indicating or recording thermometer. Keep the unit door closed as much as possible to maintain temperature.

Cut melons are a potentially hazardous food.

Important Procedures For Cold Storage

- Keep refrigerated foods at 41°F (5°C) or below and frozen foods solidly frozen during storage.
- Rotate refrigerated and frozen foods on a first in, first out (FIFO) basis and store foods in covered containers that are properly labeled and dated.
- Store foods in refrigerated and freezer storage areas at least 6 inches off the floor and space products to allow the cold air to circulate around them.
- Store raw products under cooked or ready-to-eat foods to prevent cross contamination.
- Keep different species of raw animal foods separate during storage. If limited storage space makes it necessary to store different species in the same area of the refrigerator, store poultry on the bottom shelf, ground beef and pork on the middle shelf, and fish, eggs and other cuts of red meat on the top shelf.

Freezer storage is designed to keep foods solidly frozen. Freezer equipment must also be equipped with indicating or recording thermometers to monitor the temperature of the ambient air inside the unit. If your freezer is not frost free, defrost it regularly to ensure proper operation. Wrap frozen foods and transfer them to the refrigerated storage area until the defrosting process is complete.

Although bacteria are generally not destroyed by freezing, parasites can be killed if foods are frozen at the proper temperature for the proper length of time. Guidelines have been established for destroying parasites in raw-marinated and marinated, partially cooked fish.

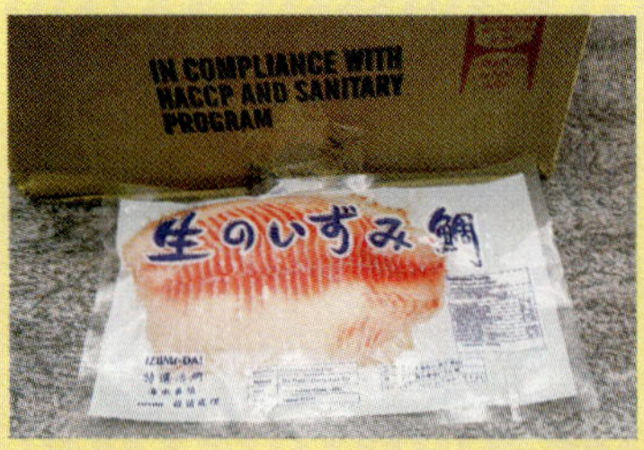

Food should be frozen throughout to -4°F (-20°C) and held for 7 days in a freezer,
or
Food should be frozen throughout to -31°F (-35°C) using a blast chiller, and held at that temperature for 15 hours.

Guidelines for parasitic destruction (Source: *2001 Food Code*)

Yellowfin, bigeye, bluefin-northern, bluefin-southern, and certain other species of tuna may be served or sold in a raw, raw-marinated, or partially cooked ready-to-eat form without freezing.

Dry Storage

Products in dry storage areas are usually packed in labeled cans, bottles, jars, and bags. The area should have a room temperature of 50°F (10°C) to 70°F (21°C) with a relative humidity of 50 to 60% to maximize shelf life of stored products. Windows should be blocked or shaded.

Dry storage area

Store dry foods on slatted shelving.

- Use slatted shelves that allow circulation of air, are at least 6 inches off the floor, and are away from the wall. This allows for cleaning under the shelving and discourages pest harborage.
- When bulk items are moved into bulk food grade containers with tight-fitting lids, include codes, labels and dates.
- Scoops and other utensils should be food grade and have long handles that keep hands from touching food.

Chemical Storage

Toxic chemicals, such as cleaners, sanitizers, and pesticides, are commonly used and sold in retail food establishments. Most of these products can be poisonous if consumed accidentally.

Many chemicals used in retail food establishments are poisonous if consumed. Others can cause irritation to skin and respiratory system.

A good label identifies the chemical, provides directions on how to use it safely, and instructs people on what first-aid measures to use in case of accidents.

- All products must be labeled and kept separate from food products. If an adequate storage area is not available, use a locked cabinet to store the chemicals.
- Identify the chemical and include directions on proper use.
- Train employees on how to use these products safely.
- It is a good practice to post lists of instructions so users can easily see when and how to use the products.

Do not store chemicals near food.

Storage Conditions for Foods

Product	Storage Conditions
Meat and meat products	• Store for up to three weeks at temperatures between 28°F (-2°C) and 32°F (0°C) and a relative humidity between 85 and 90% • Cold temperatures extend the shelf life of red meats by slowing down the growth of bacteria that cause spoilage and reduces shrink loss • Store for several months when held at 0°F (-18°C) or below • Frozen meats must be wrapped in moisture-proof paper to prevent them from drying out • Packaging for frozen foods should also be strong, flexible, and protect against light • Use by manufacturer's shelf life criteria.
Poultry	• Store at temperatures between 28°F (-2°C) and 32°F (0°C) for short periods of time • A relative humidity of 75 to 85% is recommended, as excessive humidity causes sliminess due to excessive bacterial growth • Poultry should be wrapped carefully to prevent dehydration, contamination, and loss of quality • Frozen poultry and poultry products can be stored for four to six months when held at 0°F (-18°C) or below • Use by manufacturer's shelf life criteria.
Whole shell eggs	• Keep fresh for up to two weeks when stored at 41°F (5°C) or below • It is recommended to store eggs at 34°F (1°C) to 38°F (3°C) to maintain optimum quality • Keep eggs covered and store them away from onions and other foods that have a strong odor • Discard eggs that are dirty or cracked • Always make sure to wash your hands after handling whole shell eggs • Use by manufacturer's shelf life criteria.

(cont.)

Storage Conditions (continued)

Product	Storage Conditions
Egg products (such as whole eggs, egg whites, and yolks)	• Are pasteurized to destroy *Salmonella* bacteria • Store at 41°F (5°C) or below • Store frozen eggs at 0°F (-18°C) or below and keep them frozen until time for defrosting • Once dried eggs have been reconstituted, they are considered potentially hazardous and must be stored at 41°F (5°C) or below • Use by manufacturer's shelf life criteria.
Milk	• Pasteurized milk may be held at 41°F (5°C) or less for up to ten days or longer. • The optimal storage temperature for fluid milk is 34°F (1°C) to 38°F (3°C), and the shelf life of milk is shortened significantly at higher storage temperatures. • Milk also picks up odors from other foods. Store milk in an area away from onions and other foods that give off odors. • Use by manufacturer's shelf life criteria.
Fish and shellfish	• More perishable than red meats even when refrigerated or frozen. • Fish and shellfish should be stored at temperatures ranging from 30°F to 34°F (-1 to 1°C). • Fish should be kept on crushed ice drained away from the product or solidly frozen. Recommend using fresh fish within 24 hours or less. • Shellfish shells should close when tapped. Dead shellfish must be discarded. Lobsters and clams should be kept alive until cooked or frozen. Keep shellfish tags for 90 days after purchase. If a foodborne outbreak occurs, the tags help identify the source.

(cont.)

Storage Conditions (continued)

Product	Storage Conditions
Fresh fruits and vegetables	• Require temperatures between 41°F (5°C) and 45°F (7°C) in a relative humidity of 85 to 90%. • If fruits and vegetables arrive packed in airtight film, notify your supplier to correct this issue to allow the product to respire. • Produce should not be washed before storage—wash before using. • Proper circulation is necessary to maintain freshness and firmness. Discard fruits that begin to spoil. • Whole citrus fruits and bananas should not be refrigerated.
Modified atmosphere packaging (MAP) and sous vide products	• MAP products are perishable foods and must be kept at temperatures recommended by the processor. • Most will need refrigeration at 41°F (5°C) or below. If frozen, keep solidly frozen until thawed and used. • Check expiration dates before using. Discard out-of-date products. • Do not use packages that have signs of microbial growth (slime, bubbles, molds, etc.).

Food Preparation, Handling, and Service

Preparation, handling, and service of food in retail food establishments involve one or more steps. In convenience stores, ready-to-eat products, such as sandwiches and salads, are sold to customers with minimal preparation. In larger retail food establishments, foods are prepared in consumer-sized portions using commercially available equipment for cutting, grinding, slicing, cooking, and cooling. These foods are commonly referred to as "meal solutions" or "home meal replacements."

During preparation, an important technique that can be used to promote food safety is "small batch" preparation. Food preparation is usually done at room temperature. This is several degrees into the temperature danger zone. Therefore, you must limit the amount of time the food is in the danger zone by working with small and manageable amounts of potentially hazardous ingredients. In addition, some preparation procedures may require ingredients to be pre-cooled prior to preparation (i.e., ingredients used in the preparation of chicken salad).

Ingredient Substitution

For meal solutions and home meal replacements, there is normally a recipe for the products prepared in the retail food establishment. The recipe usually includes a list of ingredients and instructions for how to prepare, store, and label the food item.

- When one or more of the original ingredients is not available for a recipe, other ingredients may be substituted so the food item can still be prepared and sold
- All ingredient substitutions should be identified in the recipe before preparation
- Ingredient substitutions must never compromise the safety of the food and should not be allowed unless they are identified and allowed in the recipe.

Avoiding Temperature Abuse

Temperature and time abuse is when food is kept in the temperature danger zone, 41°F (5°C) to 135°F (57°C), long enough for harmful organisms to grow.

- Monitoring and controlling food temperatures are extremely effective ways to minimize the risks of foodborne illnesses.
- Thermometers are used for stored, cooked, hot-held, cold-held, and reheated foods. Before using a thermometer, make sure it is clean, sanitary, and properly calibrated. Always insert the "sensor" portion or probe stem of the thermometer into the thickest part of the food. In most instances, this will be at the center of the food product or container.

Thawing

The *FDA Food Code* requires raw animal foods to be thawed in less than four hours including the time it takes for preparation for cooking or to lower the food temperature to 41°F (5°C) under refrigeration. Thawed portions of ready-to-eat foods should not be allowed to rise above 41°F (5°C) when using the cool water thawing process.

Retail food establishments will sometimes use a slacking (defrosting) process to moderate the temperature of foods prior to cooking or reheating. During the slacking process, foods can be defrosted under refrigeration that maintains the food at 41°F (5°C) or less or at any temperature if the food remains frozen. The slacking process is typically used with previously block-frozen food such as spinach.

Under no circumstances should foods be thawed or slacked at room temperature. Room temperature thawing puts foods in the temperature danger zone—the very thing you don't want to have happen. When foods are thawed at room temperature, the outer surface of the food thaws first and will soon reach room temperature. Microbial growth occurs very quickly at room temperature.

Guidelines for Thawing Food

Refrigeration

- Under refrigeration that maintains the food temperature at 41°F (5°C) or below.

Submerged under Running Water

Completely submerged under running water:

- At a water temperature of 70°F (21°C) or below
- With enough water force to remove contaminants from the surface of the food
- For a period of time that does not allow thawed portions of ready-to-eat foods to rise above 41°F (5°C)
- For a period of time that does not allow thawed portions of a raw animal food requiring cooking to be in the temperature danger zone for more than a total time of four hours.

As Part of the Cooking Process

Thawing for Immediate Service

- Use any procedure (i.e., microwave oven) that thaws a portion of frozen ready-to-eat food prepared for immediate service in response to an individual customer's order.

Cold-Holding

Most harmful microorganisms start to grow at temperatures above 41°F (5°C). Some bacteria, such as *Listeria monocytogenes*, can grow slowly at temperatures below 41°F (5°C).

Cold display of raw, potentially hazardous foods

Cold-Holding At A Glance: Refrigerators, Display Case & Cold Service Bars

Cold raw potentially hazardous (like meat)
Temperature: Below 41°F (5°C)

- Physical barriers should be in place to separate different species (meat, poultry, seafood) and to separate raw from ready-to-eat foods

Meat cutting operation
Temperature: Below 41°F (5°C)

- Slows bacterial growth
- Allows employees to work comfortably

Fish and seafood
Temperature: Below 41°F (5°C)

- Use ice from potable water
- Transport ice in "approved food-contact" containers
- Liquid must be drained from ice to prevent contamination
- Ice in contact with fish and shellfish is considered contaminated — DO NOT REUSE.
- Cooked and raw product should be kept separate.

Cold ready-to-eat potentially hazardous foods
Temperature: Below 41°F (5°C) up to 7 calendar days

- If held more than 24 hours, *FDA Food Code* recommends prepared and held products be marked with "sell by" or "use by" dates.

Ready-to-eat salads
Temperature: As cold as possible and above 32°F (0°C) and below 41°F (5°C)

- Pre-chill ingredients before using
- Prepare small batches to assure food is not in the temperature danger zone too long.

Frozen, Ready-to-eat Foods

When large amounts of food are removed from the freezer, they should be marked to indicate the date by which the food must be sold. Ready-to-eat potentially hazardous food must be used within seven calendar days or less after the food is removed from the freezer, minus the time before freezing. Subtract any time if the food is maintained at 41°F (5°C) or less before freezing.

When displaying ready-to-eat potentially hazardous foods like prepared salads and luncheon meats, be sure the refrigerator unit can maintain a safe cold-holding temperature. This may be more difficult in open top and open front refrigerated display cases that do not have doors. Refrigerated display cases have a "safe load line." This line indicates the level below which foods must be stored to ensure the food is held at the proper temperature. It is also important to store foods so the discharge or return air vents are not blocked.

Prepared salads

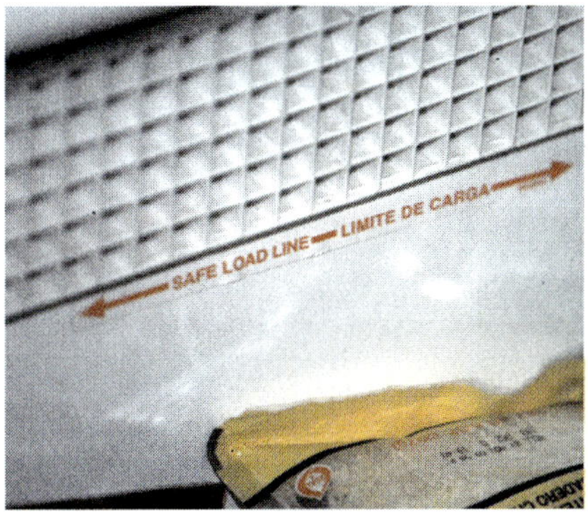

Safe load line

Cold storage of ready-to-eat foods

Prepackaged Foods

A refrigerated, ready-to-eat, potentially hazardous food prepared and held in a retail food establishment for more than 24 hours, or originating from an original container prepared and packaged by a food processing plant and opened at the retail food establishment must be discarded if it:

- Exceeds the prescribed time and temperature requirements,
- Is in a container or package that does not bear a date or day, or
- Is marked with a date or day that exceeds the time and temperature combinations described above.

Cooking

Meat, poultry, fish, seafood, eggs, and unpasteurized milk should not be prepared and served raw or rare. Establishments that choose to serve raw foods increase the risk of causing a foodborne illness. Raw animal foods need to be cooked to the proper temperatures to be safe. Many states and jurisdictions require an advisory be posted to warn consumers of the risk.

Most foods are cooked using stoves, conventional ovens, and microwave ovens. Because heat transfer can be different depending on the heating source, final temperature requirements have been set for conventional oven cooking and microwave cooking.

Cooking Guidelines for Potentially Hazardous Foods

Food Type	Minimum Internal Temperature	Minimum Time Held at Internal Temperature Before Serving
Beef Roast (rare)	130°F (54°C) 140°F (60°C)	112 min. 12 min.
Eggs, Beef and Pork (other than roasts), Fish	145°F (63°C)	15 sec.
Ground Beef, Ground Pork, and Ground Game Animals	155°F (68°C)	15 sec.
Beef Roast (medium), Pork Roast, and Ham	145°F (63°C)	4 min.
All Poultry, Stuffed meats	165°F (74°C)	15 sec.
Note: When microwave cooking, heat raw animal foods to a temperature of 165°F (74°C) in all parts of the food.		
(Source: *FDA Food Code*)		

Casseroles and other foods that contain a combination of raw ingredients such as meat and poultry must be cooked to a final temperature that coincides with the highest risk food. In this case, 165°F (74°C) is required to destroy pathogens that may be found in the poultry. Food mixtures, such as chili and beef stew, must be cooked to 165°F (74°C) to assure proper destruction of disease-causing agents.

When cooking foods in the microwave oven, the distribution of heat is often uneven. Stirring and rotating the food during the cooking process will enable heat to be distributed more evenly. The *FDA Food Code* requires raw animal foods cooked in a microwave oven to be heated to 165°F (74°C) in all parts of the food. As a common practice, foods cooked in a microwave oven should be allowed to stand covered for two minutes before serving to allow the heat inside the product to disperse more evenly.

The internal temperature of raw animal foods cooked in a microwave oven must reach 165°F (74°C) or above.

For most foods, the internal temperature will be measured by inserting the probe of a thermometer or thermocouple into the center or thickest part of the food mass. This will give you an accurate reading of the internal temperature of the product.

Cooling

Foods are in the temperature danger zone during cooling and there is no way to avoid it. After proper cooking, potentially hazardous foods need to be cooled from 135°F (57°C) to 41°F (5°C) as rapidly as possible. The *FDA Food Code* recommends hot foods, not used for immediate service or hot display, be cooled from 135°F (57°C) to 70°F (21°C) within two hours, and from 135°F (57°C) to 41°F (5°C) or less within six hours.

Improper cooling is one of the leading contributors to foodborne illness in retail food establishments.

Potentially hazardous foods prepared from ingredients, such as reconstituted foods and canned tuna, which have been held at room temperature, must be cooled to 41°F (5°C) or less within four hours.

Ice bath

Place food in shallow pans 2 to 3" deep.

Safe cooling methods

Common Methods For Reducing Cooling Time

- Blast chillers
- Walk-in coolers, loosely covered
- Use containers that facilitate heat transfer (stainless steel)
- Transfer food into shallow pans that will allow for a product depth of 3" or less
- Transfer food into smaller containers
- Place container of hot food in an ice water bath
- Stir food while cooling
- Use cooling paddles to stir the food
- Add ice as an ingredient directly to a condensed food.

Foods must pass through the temperature danger zone as quickly as possible.

Never assume any one method is working without checking the temperature and time foods take to cool. Always depend on the thermometer reading with any of the methods you use to cool food.

Always use a thermometer to verify foods are cooling properly.

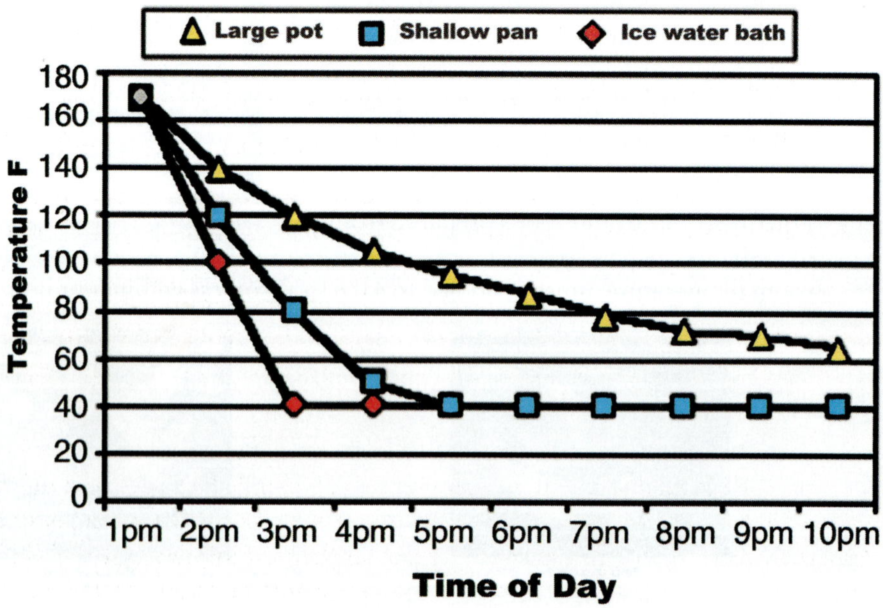

Different cooling methods for chili

Hot-Holding and Reheating

All potentially hazardous foods that have been cooked and are intended to be held hot (not cooled, stored, and reheated) must be maintained at 135°F (57°C) or above.

- Hot-holding is also required when hot potentially hazardous foods are delivered to sites away from the retail food establishment
- During hot-holding, never add fresh product to existing product and always work in small batches
- Reheat to at least 165°F (74°C) within two hours.

Hot-holding—Store hot-held foods at 135°F (57°C) or above.

While temperature is usually the most important factor in controlling microbes, there are some situations where controlling time can also be used. That is why there is an allowance of four hours in the temperature danger zone for ready-to-eat potentially hazardous food held for food service or immediate consumption. Sliced pizza and fried chicken are good examples. If these potentially hazardous foods were hot-held at 135°F (57°C) or above, the food may dry out and the quality may deteriorate very quickly.

Ready-to-eat foods commercially prepared and packaged and from a food processing plant under regulatory inspection should be free of harmful microorganisms. Therefore, these foods may be reheated to a temperature of at least 135°F (57°C) for hot-holding [rather than a minimum of 165°F (74°C)]. Commercially prepared and cooked soups would be a good example of a food that would need to be reheated to 135°F (57°C) or above followed by hot-holding.

Avoid holding potentially hazardous foods in the temperature danger zone. Most establishments will limit this amount of time to 20-30 minutes. In these instances, it is critical to achieve the required cooking temperature and the amount of time foods are held in the temperature danger zone must be carefully monitored and recorded. The procedures and monitoring required when using time as a method of control varies. Consult with your local regulatory authority for requirements in your jurisdiction.

Food Handling

Employees must practice good personal hygiene when handling food. This starts with a clean uniform and an effective hair restraint.

- Food handlers in all departments should avoid touching food with their bare hands
- They can use tongs, serving spoons, disposable gloves, or deli tissue when handling meats, cheeses, prepared salads, or when making sandwiches
- Employees must hold serving utensils by the handle only, and they must never touch the part of the utensil that comes into contact with food
- A single utensil should be used for each food item, and the utensil should be stored in the food between uses
- Always store serving utensils in a way that permits the employee to grab the handle without touching the food.

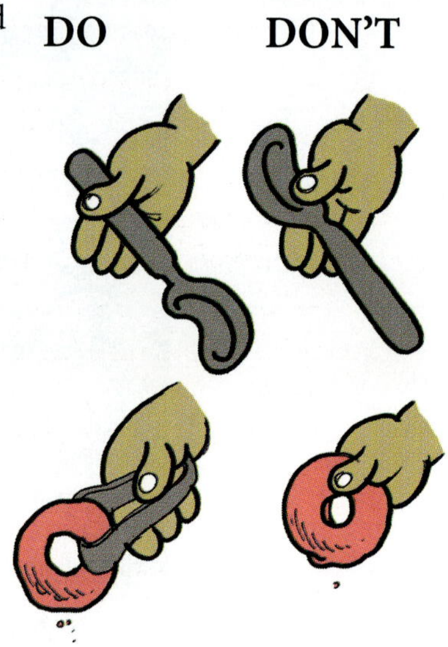

Handle eating utensils properly.

Discarding or Reconditioning Food

A food that is unsafe, adulterated or not honestly presented shall be reworked or reconditioned using a procedure approved by the regulatory authority in the jurisdiction, or it must be discarded.

Food must be discarded if it is not from an approved source or has been contaminated by food employees, consumers, or other persons via soiled hands, bodily discharges (i.e., coughs and sneezes), or other means.

Ready-to-eat foods must be discarded if they have been contaminated by an employee who has been restricted or excluded as described in Appendix D of this book.

Refilling Returnable Containers

A take-home food container returned to a retail food establishment may not be refilled with a potentially hazardous food at the establishment. A food-specific container for beverages may be refilled at a food establishment if:

- Only a beverage that is not a potentially hazardous food is dispensed into the container
- The design of the container and of the rinsing equipment and the nature of the beverage, when considered together, allow effective cleaning of the container at home or in the retail food establishment
- Facilities for rinsing before refilling returned containers with fresh, hot water that is under pressure and not recirculated are provided as part of the dispensing system
- The consumer-owned container returned to the retail food establishment for refilling is refilled for sale or service only to the same consumer
- The container is refilled by an employee of the retail food establishment or the owner of the container if the beverage system includes a contamination-free transfer process that cannot be bypassed by the container owner.

Personal take-out beverage containers, such as thermally insulated bottles, nonspill coffee cups, and promotional beverage glasses, may be refilled by employees or the consumer if the refilling process will protect the food and food-contact surface of the container from contamination.

Self-service Bar

Self-service salad and hot food buffet bars are very popular in retail food establishments. They offer convenience and a wide range of selections for customers. The most important food safety goals for this type of operation are to:

- Protect foods from contamination by customers
- Keep foods out of the temperature danger zone.

Self-service bar

Rules for Self-service Bars:

- A properly installed sneeze guard protects the food from contamination by your customers.
- Never place oysters and other types of raw animal foods on a

self-service bar except for ready-to-eat foods like sushi. Keep hot-held potentially hazardous foods at 135°F (57°C) or above and cold foods at 41°F (5°C) or below.

- Use clean and sanitized utensils in a self-service bar and replace any utensils that become contaminated or soiled.
- Use only one utensil for each food item and store it in the food between uses.

If customers are allowed to visit the self-service bar more than once, they must be given a clean plate or bowl for each trip. This will reduce the risk of contaminating food on display at the bar. Beverage cups and glasses may be reused to get refills.

Sneeze guards help prevent the spread of foodborne illness.

Checkout and Bagging

The last step in the flow of food in retail food establishments takes place at the checkout counter. Food employees must practice good personal hygiene and not allow their hands to transfer bacteria from raw animal foods to cooked and ready-to-eat foods. Juices from raw meat, poultry, fish and shellfish can contaminate produce and other ready-to-eat foods. These juices must be routinely wiped off conveyor belts, and employees must wash hands properly.

Keep Checkout Counters Clean and Sanitary

Baggers must be careful not to put raw meat, poultry and fish in the same bags with fresh fruits and vegetables. In addition, non-food items, such as detergents, pesticides, and light bulbs, must not be put in the same bag with foods to avoid

Checkers and baggers must work carefully to prevent contamination and cross contamination.

contamination with chemical and physical hazards. Freezer bags should also be used to separate frozen goods.

Sometimes customers change their mind about purchasing a particular item. Whenever these are potentially hazardous foods, the product must be put back into the proper storage or display environment as quickly as possible.

Temporary Facilities and Mobile Food Facilities

A temporary food establishment (TFE) is defined by the *FDA Food Code* as a food establishment that operates for a period of no more than 14 consecutive days in conjunction with a single event or celebration. TFEs may operate either indoors or outdoors and often have limited physical and sanitary facilities available.

Some food establishments are offering food by way of temporary and mobile facilities.A variety of foods can be prepared and served from temporary stands and trailers. Some examples include sandwiches, pizza, BBQ ribs, gyros, corn on the cob, confections (i.e., cotton candy, funnel cakes, or elephant ears), and beverages.

Mobile facilities are trucks and trailers used to cater events located away from the establishment. The extent of food items offered through catering operations is practically unlimited.

Mobile food facility

Temporary food facility

The same food safety practices employed in other areas of the establishment must also be applied at temporary, mobile facilities, and in-store product demonstrations. In particular, food must be protected from:

- Temperature abuse
- Infected employees who practice poor personal hygiene and use improper food-handling practices
- Contamination and cross contamination.

Employees must not eat and smoke around food, and they must wash their hands whenever they are contaminated.

Employees shall use tongs, utensils, and deli tissue to avoid bare hand contact with food. Disposable gloves can provide an additional barrier against contamination. However, gloves must not be viewed as a substitute for proper hand washing.

Food must be protected from contamination and cross contamination. Food on display must be protected from contamination by customers and employees. Food equipment must be designed and constructed to make it smooth, easily cleanable, nontoxic and nonabsorbent.

Use disposable utensils whenever possible. When it is necessary to use multiple-use utensils, they must be washed and sanitized using a four-step process:

1. Wash in hot, soapy water.
2. Rinse in clean water.
3. Use chemical sanitizing rinse.
4. Air-dry.

Garbage and paper wastes should be placed in containers lined with plastic bags and equipped with tight-fitting lids. Food and wastes must be kept covered to avoid attracting insects, rodents, and other pests.

Vending Machines

A vending machine is a self-service device which dispenses individually-sized servings of food and beverages after a customer inserts a coin, paper currency, token, card, key, or makes a payment by another means. Some vending machines dispense food and beverages in bulk while others dispense products in individually wrapped packages. Vending machines are available that will dispense potentially hazardous, ready-to-eat foods (i.e., sandwiches, french fries, dairy products, and soup) and nonpotentially hazardous foods (candy, snack chips, pastries, coffee and soft drinks).

Vending machines that store and dispense potentially hazardous foods must have adequate refrigeration and/or heating units, insulation, and controls to keep cold foods cold and hot foods hot. These machines must also be equipped with an automatic control that prevents the machine from vending food if there is a power failure, mechanical failure or other condition that prevents cold food from being maintained at 41°F (5°C) or below and hot food maintained at 135°F (57°C) or above.

The temperature cutoff requirement does not apply:

- In a cold food machine during a period not to exceed 30 minutes immediately after the machine is filled, serviced or restocked; or
- In a hot food machine during a period not to exceed 120 minutes immediately after the machine is filled, serviced or restocked.

Refrigerated, ready-to-eat, potentially hazardous foods prepared in a food establishment and dispensed through a vending machine (with an automatic shutoff control) shall be discarded if it a) exceeds the time and temperature combinations prescribed by the *FDA Food Code* or b) is not correctly date labeled.

Home Meal Replacement

Home meal replacements and meal solutions are the terms most often used to refer to high-quality meals prepared away from home but eaten at home. These types of products are now a multi-billion dollar business for retail food establishments throughout the country. The

Home meal replacements come in "ready-to-cook," "ready-to-heat," and "ready-to-eat" varieties.

move toward prepared food has retail food establishments selling complete meals instead of just ingredients. Retail food establishments and restaurants are now competing head-to-head for the "heat and eat" business that has become very popular with today's customers.

It is a good idea to establish sell-by or best if used by dates and codes for these ready-to-eat foods.

All three varieties are designed to save time and effort for families that are too tired to cook at the end of the day.

Ready-to-eat foods are a common variety of home meal replacements.

Home replacement meals should be labeled so customers understand how to keep the product safe when they take it home. Warn them against keeping the food in the car while they do more shopping or keeping it at room temperature when they get home. Pamphlets and brochures stuffed into bags or stapled to the front of bags can help educate customers about safe handling of home replacement meals. Retailers must be able to show they've done everything in their power, including written documentation and following industry standards, to make the food safe.

Check It Out

Back to the Story… *In this case, the best solution is to discard the éclairs because they are not labeled properly and it is unknown how long they were left at room temperature. Proper labeling using the first-in, first-out method is an important way to monitor and identify foods as they pass through the food product flow. The éclairs are potentially hazardous because they contain eggs in the custard and if stored improperly could lead to foodborne illness.* Salmonella *or* Staphylococcus aureus *microbes thrive in this type of custard. The management at this establishment should set up training programs so employees understand the importance of labeling food products during the flow of food.*

✓ Not all retail food establishment managers will actually purchase food products. However, knowledge of the rules, regulations, and procedures for receiving and storing food is a must for everyone responsible for food safety.

✓ The flow of food must be monitored and considered when preparing safe food.

✓ Foods should only be allowed in the temperature danger zone for a short time during thawing, heating, and cooling activities. The three main contributors to foodborne illness in retail food establishments are:

- Time and temperature abuse
- Cross contamination
- Poor personal health and hygiene practices by food handlers.

These factors must be controlled throughout the flow of food to assure food safety. Preparation and service are critical processes in your establishment because they are the last steps you take before your customer eats the food.

Quiz 4 (Multiple Choice)

Please choose the BEST answer.

1. The **safest** way to thaw food is:
   a. At room temperature.
   b. In the refrigerator.
   c. Under ultra-violet light.
   d. None of the above.
2. What is the minimum number of days shellfish tags must be kept after the product is sold?
   a. 15.
   b. 30.
   c. 60.
   d. 90.
3. 3.Which of the following potentially hazardous foods can be received at 45°F (5°C) or below rather than 41°F (5°C) or below?
   a. Milk.
   b. Liquid eggs.
   c. Fish.
   d. Ground lamb.
4. Foods that are held hot, like soup, should be at or above temperatures of:
   a. 165°F (74°C).
   b. 155°F (68°C).
   c. 145°F (63°C).
   d. 135°F (57°C).

5. Temperatures in refrigerated storage should be measured daily with the thermometer placed in the part of the unit that is:

   a. Coldest.

   b. Warmest.

   c. Highest.

   d. Lowest.

Answers to the multiple-choice questions are provided in Appendix A.

References/Suggested Readings

Food and Drug Administration (2001). *2001 Food Code.* U.S. Public Health Service. Washington, DC.

Food and Drug Administration (2001). *Fish and Fishery Products Hazards and Control Guide - Third Edition.* U.S. Public Health Service. Washington, DC.

Longrèe, Karla, and G. Armbruster (1996). *Quantity Food Sanitation.* John Wiley and Sons, Inc. New York, NY.

Thayer, David W., et al. (1996). *Radiation Pasteurization of Food. Council for Agricultural Science and Technology Issue Paper, No. 7.* April, 1996

Learn How To:

- Understand the importance of properly maintaining equipment and utensils and the influence it has on food safety.
- Describe how work tasks are conducted in work centers.
- Understand how the preparation and service of food flows through retail food establishments.
- Understand the basic design requirements that apply to floor- and table-mounted equipment used in retail food establishments.
- Recognize the different types of cooking, refrigeration, preparation and dishwashing equipment available for use in retail food establishments.
- Describe how proper installation and maintenance affect the operation of equipment used during the course of food production, holding, display, and handling.
- Explain the role of proper lighting in food production and warewashing areas.
- Explain how proper heating, air conditioning, and ventilation affect food sanitation and employee comfort and productivity in retail food establishments.

CHAPTER 5B

Facilities, Equipment, and Utensils

Proper Equipment Usage . . .

The local health inspector arrived for a routine visit. The manager on duty informed the staff of the inspection and joined the inspector on his review of the establishment. The inspection consisted of a walk through every department checking for critical and noncritical violations. During the inspection, they walked past the upright display case in the dairy department. The inspector noted the cartons of eggs were stacked 10 inches above the load line. The manager immediately called an employee over and told him to take care of the problem.

What is the problem, what did the manager do correctly, and how should the employee correct it?

Design, Layout, and Facilities

The design, layout, and facilities provided in a retail food establishment must be consistent with the types of foods being sold there. The equipment used in the various departments must be properly designed and sized to meet the production needs of the area. A layout that works well in one establishment may not necessarily be suitable for another that produces and sells different food items. Most equipment and facilities will be used for several years. Therefore, managers should seek a design that meets current needs and accommodates increased sales and new items that are offered for sale in the future. Some general areas commonly found in retail food establishments are:

- Delivery and receiving
- Storage
- Preparation
- Hot- and cold-holding
- Display
- Handling and service
- Warewashing, cleaning, and sanitizing
- Waste storage and pickup
- Housekeeping
- Toilet facilities.

Display area

It is important you develop a flow diagram of your establishment's operation in order to plan the physical facilities for each function.

When planning a retail food establishment:

- Understand and visualize each function that will be conducted within the different departments
- Determine the specific tasks employees must complete when performing the function
- Arrange the tasks in a way that allows a smooth and sequential flow within that department or area.

The activities carried out by employees in the various departments are called functions.

Regulatory Considerations

When planning facilities for retail food establishments, you must know about and comply with national, state, and local standards and codes related to:

- Health
- Safety
- Building
- Fire
- Zoning
- Environmental code standards.

The **equipment** used in retail food establishments should meet the American National Standards Institute (ANSI) standards and bear the stamp of approval of recognized third-party certification organizations. Some examples of these organizations include:

- NSF International
- Underwriters Laboratories, Inc. (UL)
- American Gas Association (AGA).

Buyers who purchase equipment approved by these organizations are assured quality materials are used in the construction. Also, the items are designed and constructed to meet accepted food sanitation criteria.

UL and NSF seals

Work Center Planning

The departments in a retail food establishment are commonly organized into work centers. These are areas where a group of closely related tasks are performed by an individual or individuals. The number of work centers required in a department depends on the number of functions to be performed and the volume of material to be handled.

An employee should be able to complete the related tasks at the work center without moving away from it. The work center should also be large enough to do the job yet small enough to reduce travel and conserve time and effort. A properly designed work center will provide adequate facilities and space for:

- Efficient production
- Fast handling and service
- A pleasant environment
- Effective cleanup.

Plan work centers carefully.

Equipment Selection

It is extremely important to select the right piece of equipment for the job. Compare different pieces of equipment for a particular job and look at such features as:

- Design
- Construction
- Durability
- Ability to clean easily
- Size
- Cost
- Safety
- Overall ability to do the job.

Purchase equipment that will improve the quality of food, reduce labor and material costs, improve sanitation, and contribute to the bottom line of the establishment.

Size and Design

Design is an important feature of food equipment, since the equipment used in a retail food establishment is subject to constant use and abuse. Equipment and utensils must be designed to function properly when used for their intended purposes.

Purchase equipment that fits into the space available and can handle your anticipated future needs.

Equipment that sits on the floor must be:

- Elevated on 6-inch (15 cm) legs, or
- Sealed to the floor, or
- Mounted on casters to make it easily movable.

Clearance space and mobility make it easier to clean the floor under and behind the equipment. Equipment sealed to the floor will prevent the accumulation of debris and the harborage of pests.

Floor-mounted equipment

Table-mounted equipment

In retail food establishments, display shelving units, display refrigeration units, and display freezer units located in the customer shopping areas do not have to be elevated or sealed to the floor. However, the floor under these units must be maintained in a clean and sanitary manner.

Table-mounted equipment (that is not easily movable) should be on 4-inch (10 cm) legs. This provides clearance between the counter top and the bottom of the equipment and makes it easier to clean under and around the equipment. The equipment may also be sealed to the counter.

Major costs associated with the purchase of equipment:

- Purchase price
- Installation cost
- Operating costs
- Maintenance costs
- Finance charges.

Construction Materials

The *FDA Food Code* and construction standards, such as those from ANSI, require food equipment and utensils to:

- Be smooth
- Be seamless
- Be easily cleanable
- Be easy to take apart
- Be easy to reassemble
- Have rounded corners and edges.

Materials used in the construction of utensils and food-contact surfaces of equipment must be nontoxic and not impart colors, odors, or tastes to foods. Under normal use, these materials must also be safe, durable, corrosion-resistant, and resistant to chipping, pitting, and deterioration.

Food-contact surfaces are the parts of equipment and utensils that normally come into contact with food or from which food may drain, drip, splash, or spill into food or onto a surface that is normally in contact with food.

Non-food-contact surfaces are the remaining parts of the equipment and utensils and the surrounding area that should not make contact with food during production.

Metals

Metals are very popular materials in retail food establishments. Chromium over steel gives an easily cleanable, high-luster finish. It is commonly used in conjunction with small appliances. Noncorrosive metals formed by the alloys of iron, nickel, and chromium may also be used in the construction of food equipment.

Lead, brass, copper, cadmium, and galvanized metal can cause a chemical poisoning when they come into contact with high-acid foods (foods that have a low pH). Therefore, these materials must not be used as food-contact surfaces for equipment, utensils, and containers.

Abrasive cleaners and scouring pads can scratch the surface of the metal and cause harborage areas for disease-causing microorganisms.

Stainless Steel

Stainless steel is one of the most popular materials in retail food establishments. It is commonly the material of choice for food containers, table tops, sinks, dish tables, dishwashers and ventilation hood systems. Stainless steel has a durable, shiny surface that easily shows soil and is easy to clean and maintain. Stainless steel also resists high temperatures, rust, and stain formation.

One factor that influences the cost of stainless steel is the extent of polishing desired, because polishing requires labor, material, and energy. Finish No. 4 (on a scale of 1-8) is most commonly preferred for food production areas whereas higher finishes are usually preferred in display and handling areas.

Plastic

You must be sure to buy food equipment that is made of only food-grade plastics. Select the one that works best for you based on intended use and durability. The harder, more durable plastics are easier to clean and sanitize. Some examples of plastics used in retail food establishments are:

- Acrylics (used to make covers for food containers)
- Fiberglass (used in boxes, bus trays, and trays)
- Polyethylene (used in storage containers and bowls).

Plastic and fiberglass are very popular in retail food establishments because they are durable, inexpensive, and can be molded into different combinations.

Wood

The *FDA Food Code* permits limited use of wood materials including hard maple or an equally hard, close-grained wood for cutting boards, cutting blocks, and baker's tables. Wood is also approved for paddles used in pizza operations.

Advantages of Wood:	Disadvantages of Wood:
• Light in weight • Economical.	• Porous to bacteria and moisture • Absorbs food odors and stains • Wears easily under normal use • Requires frequent maintenance and replacement.

The consensus seems to be the disadvantages of using wood for food-contact surfaces outweigh the advantages. This is primarily because of the high degree of maintenance required and problems with keeping the equipment clean and sanitary.

Types of Equipment

NSF International and Underwriters Laboratories, Inc. provide independent evaluations of equipment and materials. The NSF and UL emblems on a piece of equipment verify the equipment has been tested and meets the requirements prescribed in their standards.

Cooking Equipment

The most important criteria to use when selecting cooking equipment are the types and quantities of food prepared, ease of cleaning, durability and energy conservation. The frame, door, exterior, and interior materials of cooking equipment should contribute to the durability and cleanability of the equipment. The type and thickness of insulation materials contribute to the energy efficiency of this equipment.

Ovens

Ovens are important pieces of equipment in the bakery and may well be used by other departments in a retail food establishment. The heat in an oven is distributed by radiation, conduction, or convection, depending on the type oven being used. A good oven should rise to 450°F (232°C) within 20 minutes, and proper heat circulation is important. Ovens should be able to cool quickly when a drop in temperature is required. All ovens should be well insulated to prevent heat loss. Ovens should be in a well-ventilated area.

Range 	**Primary Department: Bakery** **Heat Distribution: Conduction** • Commonly used in small operations • Cooking surface on top of oven.
Rotisserie 	**Primary Department: Deli** **Heat Distribution: Convection** • Used to cook and hold poultry and meats • Powered by gas, electricity, or wood burning.

Deck 	**Primary Departments: Bakery, Deli** **Heat Distribution: Conduction** • Multiple ovens stacked on top of one another • Each oven contains separate heating elements.
Convection 	**Primary Departments: Bakery, C-Store** **Heat Distribution: Convection** • High speed fan circulates heated air around food to reduce cooking time • Multiple racks allow for more cooking in a smaller space.
Microwave 	**Primary Department: C-Store** **Heat Distribution: Radiation** • Used for thawing, heating and reheating foods • Cooks small quantities of food quickly.

Other types of ovens may be used in retail food establishments. These include rotary, infrared, conveyor, and roll-in units. Each of these pieces of equipment has unique features and has been designed for special applications. You should contact your equipment supplier to determine if this type equipment is best suited to carry out the functions required in your operation.

Refrigeration and Low-temperature Storage Equipment

Refrigerators and freezers are used to keep perishable foods fresh and preserve the safety and wholesomeness of potentially hazardous foods. Many foods deteriorate rapidly at room temperature. Retail food establishments can reduce spoilage, waste, and shrinkage by keeping foods at lower temperatures until they are used.

All types of refrigerators and freezers have a maximum capacity for cooling foods. Too often, employees put large amounts of hot food in a unit and rely on it to "cool" the food. On the contrary, the addition of large amounts of hot food causes the inside temperature of the unit to rise above acceptable storage temperatures. This, in turn, causes food in the unit to be stored in the temperature danger zone. Don't forget the primary purpose of a refrigerator is to take foods that are already at or near the proper cold-holding temperature and keep them out of the temperature danger zone.

The efficient operation of a refrigeration unit depends on several factors including:

- Design
- Construction
- Capacity of the equipment.

Proper air circulation both inside and outside the refrigerator makes it more efficient.

The storage of food in shallow containers, placed on slatted shelves or tray slides to permit good circulation of the chilled air is essential for both short- and long-term storage. Do not line shelves in refrigerators and freezers with sheet pans, foil, plastic, or cardboard. This decreases airflow in the storage compartment and reduces the cooling capacity of the equipment. Proper construction of cold storage units includes sturdy construction of doors, hardware, and fixtures. Doors may be full-length or half-length, and shelving should be adjustable. Door gaskets should be durable and easy to clean. They must be replaced when worn.

The size of refrigerator or freezer needed depends on the size of the work area and the type and amount of food to be stored in the unit. Refrigeration and low-temperature storage equipment must be adequately sized and properly installed to assure reliable and efficient operations.

Refrigeration equipment

Maintenance of the refrigeration equipment in the various departments is an important responsibility. These units must be cleaned on a regular basis to maintain good sanitary conditions and eliminate odors. The inside walls, floor, shelves and other accessories of a walk-in refrigeration unit must be cleaned regularly to remove spills and debris. Don't forget to clean the fan grates and condenser as part of your routine cleaning.

Reach-in Refrigeration

Reach-in

(Courtesy of Hobart Corp.)

(Courtesy of Hobart Corp.)

Primary Departments:

- **Bakery**
- **Deli**
- **C-Store**

- Models range in capacity
- Can have multiple doors
- May have external thermometer attached to a sensing device inside the warmest part of the refrigeration unit.

Walk-in

Primary Departments:

- **Dairy**
- **Meat**
- **Seafood**
- **Deli**
- **Produce**
- **C-Store**

- Stores large quantities of perishable foods between 32°F (0°C) and 41°F (5°C)
- Used to thaw products
- Can be combined with reach-in dairy/deli display cases that are loaded from inside the walk-in
- Door openings may have 4" wide plastic strips called strip curtains which reduce cool air loss when the door is open
- May have external thermometer attached to a sensing device placed at the warmest location in the refrigeration unit.

Display

Primary Departments:

- Grocery
- Meat
- Seafood
- Deli
- Dairy
- Produce
- C-Store

- Keep cold foods out of the temperature danger zone and frozen foods solidly frozen while on display
- Avoid stocking above the maximum load line
- Avoid covering vents and return air openings
- Properly control defrost cycles
- May be open air or have closing doors.

Cook-chill and Rapid-chill Systems

Cook-chill is a system in which food is:

- Cooked using conventional cooking methods
- Rapidly chilled using a chiller
- Stored for a limited time
- Reheated before service to the customer.

The food is typically chilled to 37°F (3°C) in 90 minutes or less and is stored at temperatures between 33°F (1°C) and 38°F (3°C) for five days. Day 1 is considered the day of production and Day 5 is the day of service.

The advantages of this system include reduction of peaks and valleys in production and readily available foods. The disadvantages of this system include the high cost of the chiller and the storage space requirements.

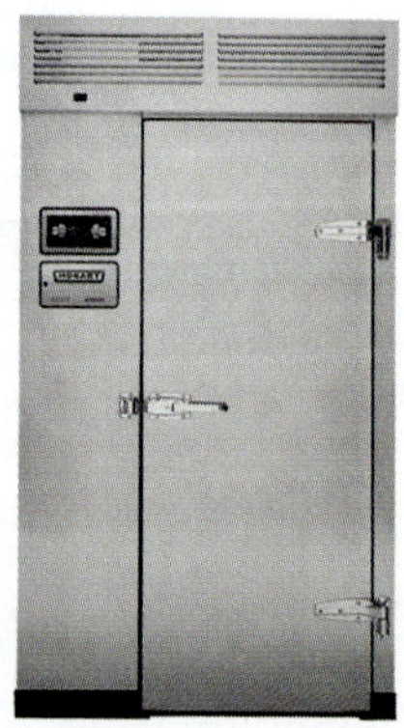

Rapid-chill Equipment

(Courtesy of Hobart Corp.)

Rapid-chill systems are designed to cool hot foods very quickly. This type equipment can typically get a few hundred pounds of hot food through the temperature danger zone in two hours or less. Although this equipment is somewhat expensive, it can be a very good investment for some retail food establishments that work with large masses of food or with foods that are challenging to cool quickly.

Watch out for metal fasteners on plastic pouches. They are a physical hazard and should be discarded immediately upon removal.

Be prepared in case of a power failure:

- 1. Keep refrigerator doors closed.
- 2. Monitor product temperatures.
- 3. Discard products that are in the temperature danger zone for more than four hours.

Hot-Holding Equipment

Potentially hazardous food that has been cooked and is to be eaten hot must be held at 135°F (57°C) or above. An important fact to remember is hot-holding equipment will not raise the temperature of foods very much. In order for hot-holding equipment to work properly, the food must be at 135°F (57°C) or above when it is put into or onto this equipment.

Hot-holding equipment

Hot-holding equipment uses steam, heating elements, or light bulbs to keep foods hot. This equipment should be checked regularly to make sure it is working properly. Food temperatures must be monitored to assure they are being maintained at 135°F (57°C) or above. Hot-holding units must be cleaned regularly, and food should be rotated using a FIFO procedure.

Other Types of Food Equipment

Slicers

The basic design of food slicers includes a circular knife blade and carriage that passes under the blade. Foods to be sliced are placed on the carriage and fed either automatically or by hand. Slicers can be dangerous if not used properly. The manufacturer's instructions should always be followed, and employees should receive training on safe operation and proper cleaning of this equipment.

Slicer

(Courtesy of Hobart Corp.)

Mixers, Grinders, Choppers, Tenderizers and Saws

Mixers, grinders, choppers, tenderizers, and saws are examples of equipment commonly found in a retail food establishment. Some of these pieces of equipment are available as table and floor models.

Mixers can also be used to shred and grind when different accessories and attachments are used. Floor model mixers have three standard attachments: (1) a paddle beater for general mixing which can be used to mash, mix, or blend foods and ingredients; (2) a whip to incorporate air into products; and (3) a dough hook used to mix and knead dough.

Mechanical tenderizers are commonly used to improve the edible quality of meat. Tenderizers cut connective tissue in meats and poultry by making incisions with stainless steel blades. The depth and the spacing of the blades and the frequency with which the meat is penetrated control the amount of tenderizing.

Grinders and choppers work well with a variety of fresh foods and ingredients including meats and vegetables.

Band saws are commonly used in meat and fish processing areas. Heavy-duty saws are best suited for large volume retail food establishments. These saws are designed to cut meat and bone and hold up under continuous heavy use. Table top and light-duty saws are commonly used in small delis, meat shops, and small establishments. This equipment is excellent when cutting boneless meats, chicken, and fish.

Meat grinder

(Courtesy of Hobart Corp.)

Band saws must be durable, easily cleanable and safe to use. Food-contact surfaces must be made of nontoxic materials that provide good sanitation qualities. Employees must be able to disassemble this equipment for easy cleaning, preferably without tools. Band saws must be properly guarded or have a "dead man" switch to protect the operator from contact with the blade.

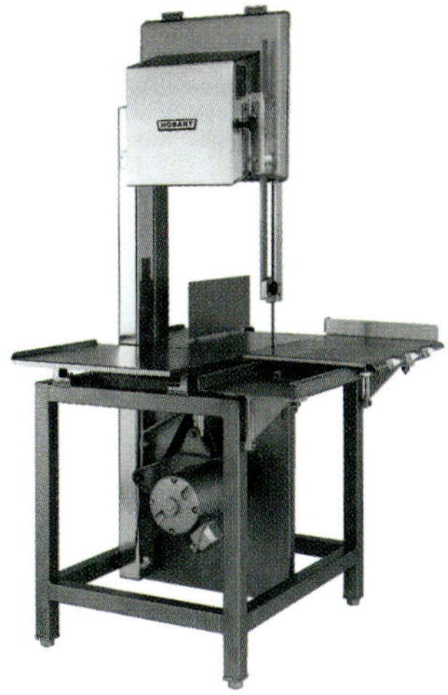

Band saw

(Courtesy of Hobart Corp.)

Ice Machines

Ice is an important item in retail food establishments. It is used to chill beverages and preserve the freshness of fish and some produce items during display.

Ice must be made from potable water, and ice machines must protect the ice during production and storage.

Ice is food and must be handled with the same degree of care as other food items.

The parts of an ice machine that come into contact with the ice must be smooth, durable, easily cleanable and constructed of nontoxic materials. These food-contact surfaces must be cleaned and sanitized regularly to prevent the growth of mold and other microorganisms. The drain line from the ice machine must be equipped with an air gap to protect the ice from contamination due to backflow. You will learn more about backflow and air gaps in Chapter 7.

Scoops, shovels, carts, and other equipment used to dispense or transport ice must meet the design and construction criteria for food-contact surfaces. When not in use, this equipment should be stored in a manner that will protect it from contamination.

Employees must never dispense ice by passing a glass or cup through the ice. This can cause glass from the container to break off and become a physical hazard in the ice. Food and beverage containers must not be stored in ice that will be used for drinking purposes. This prevents contamination of the ice your customers may consume. When scoops are stored inside an ice machine, they must be placed in a bracket mounted to a wall of the ice storage compartment. This will permit

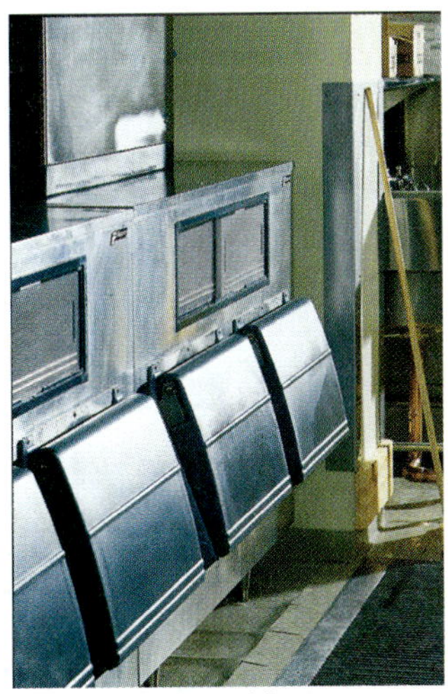

Ice machine

employees to use the handle of the scoop without their hands touching the ice or the food-contact surface of the scoop.

Do not store food and beverage containers in ice served to customers

Ambient Temperature Display Equipment

Many foods in retail food establishments are displayed at room temperature. These include shelf-stable products, certain types of produce, bakery items that have a low water activity (bagels, cookies, doughnuts, etc.) and bulk foods such as candy, peanuts, and cereals. These are not considered potentially hazardous foods, and they can be held safely at room temperature. However, they must be protected from contamination by customers and the environment.

Ambient temperature display equipment

Ambient display equipment can have both food-contact and non-food-contact surfaces. The food-contact surfaces of the equipment must be cleaned and sanitized regularly to remove contaminants. Non-food-contact surfaces must be cleaned regularly to remove spills and eliminate soil.

Stock on display should be rotated using the FIFO method, and utensils should be provided so customers will not contaminate unpackaged food with their bare hands.

Some produce display equipment is equipped with misters that help assure the freshness of fruits and vegetables and prevent wilting, dehydration and shrink. These misters must be connected to a potable water supply, and the system and heads of the system must be properly cleaned and maintained to assure maximum freshness, safety, and wholesomeness of produce exposed to the mist.

Produce case with misting heads

Live Seafood Display and Holding Tanks

Fish, crustacean shellfish (crabs and oysters), and molluscan shellfish (oysters, clams, and mussels in the shell) are live and perishable products that must be protected to remain safe, wholesome, and attractive to customers. The quality of the product in live seafood tanks is only as good as the water in the tank. Water quality is critical, and the key to maintaining optimum water quality is filtration.

Some important handling and maintenance steps to use with this type of equipment are:

- Clean and replace the ultraviolet light to assure proper intensity to kill pathogens
- Cull out and discard dead, cracked and weak molluscan shellfish daily
- Before adding molluscan shellfish to a tank, make sure they are cleaned thoroughly and cull out dead, cracked and weak shellfish
- Never mix molluscan shellfish with crustacean shellfish
- Never allow molluscan shellfish to be submerged in the system
- Never cook dead or dying shellfish from live displays as a way to reduce shrink
- Clean the tank interior to keep it free of algae and prevent slime buildup that may harbor bacterial pathogens.

Live seafood display system

Single-service and Single-use Articles

Single-service articles include tableware, carryout utensils, and other items such as bags, containers, stirrers, straws and wrappers designed and constructed to be used only one time by only one person. After one use the article is discarded.

Single-use articles include items such as wax paper, butcher paper, deli paper, plastic wrap, and certain types of food containers that are designed to be used once and discarded.

A retail food establishment should provide single-use and single-service articles for food handlers if it does not have proper facilities for cleaning and sanitizing multi-use kitchenware and tableware (see Chapter 7). These establishments must also provide single-service articles for use by consumers.

Materials used to make single-service and single-use articles must not permit the transfer of harmful substances or pass on colors, odors, or tastes to food. These materials must be safe and clean when used in retail food establishments.

Warewashing Equipment

Warewashing is the process used to clean and sanitize the equipment, utensils, dishes, glasses, etc., that are used during the preparation, handling and consumption of foods. Proper warewashing is one of the most important jobs in a retail food establishment. As you learned in Chapter 3, contaminated equipment and utensils have been identified as sources of contamination and cross contamination that can cause foodborne illness.

Equipment and utensil warewashing should be performed in a room or area separate from food production areas.

Warewashing areas must be well lighted and well ventilated. Noise-absorbing materials may be installed on walls and the ceiling to lower noise levels in warewashing areas.

Some of the items most frequently washed and sanitized in a retail food establishment are:

- **Utensils** such as knives, forks, spoons and tongs
- **Kitchenware** such as pots, pans, cutting boards, slicers, grinders, and mixers
- **Tableware** such as dishes, glasses, and eating utensils.

The purpose of warewashing is to clean and sanitize equipment and utensils. It consists of two phases:

- A cleaning phase where visible soil is removed from the surface of the item through washing and rinsing

- A sanitizing phase where the number of disease-causing microorganisms on a cleaned surface is reduced to safe levels.

Cleaning and sanitizing operations can be performed either manually or mechanically. Most retail food establishments use manual warewashing operations to clean and sanitize equipment and utensils. Some larger establishments may use a mechanical dishwashing machine to perform these tasks.

Manual Warewashing

A manual warewashing area must provide adequate space to store soiled equipment and utensils. Items to be cleaned must be pre-flushed or pre-scraped and, if necessary, pre-soaked to remove food particles and soil. A hose and nozzle or other device must be provided to pre-flush and pre-scrape food soil into a garbage container or disposal. This equipment must be located at the soiled end of the warewashing operation to avoid contaminating cleaned and sanitized equipment and utensils.

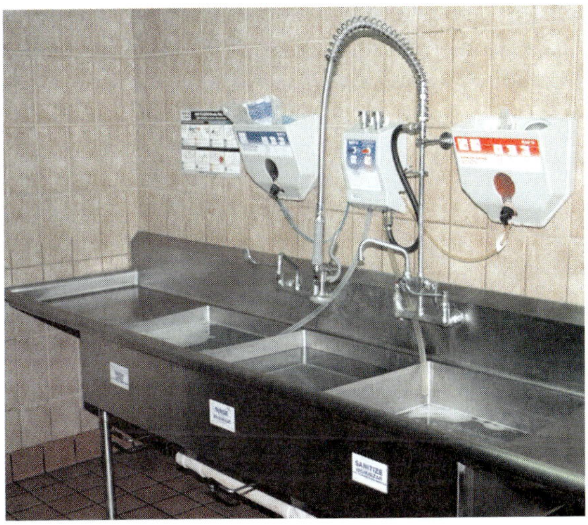

Three-compartment Sink

Manual warewashing can also be performed using a bucket and brush, wall-mounted hose units, and spray units. These processes will be explained in more detail in Chapter 6 of this book.

The compartments of the three-compartment sink must be large enough to accommodate the largest pieces of equipment and utensils used in the retail food establishment. Supply each compartment with hot and cold potable running water. Provide drainboards or easily movable dish tables of adequate size for proper handling of soiled utensils prior to washing, and for air-drying cleaned and sanitized items.

Mechanical Warewashing

Mechanical warewashing machines may be used by retail food establishments but are far less common than a three-compartment sink. Dishwashing machines can be used to clean and sanitize equipment and utensils that do not have electrical parts and that will fit into the machine. When using a single-tank, stationary-rack dishwashing machine, the dishes are placed on racks and washed one rack at a time with jets of water within a single tank. They are operated by opening a door, inserting a rack of dishes, closing the door, and starting the machine.

The most common types of mechanical dishwashers used in retail food establishments are

- **Single-tank**
- **Stationary-rack**
- **Low-temperature.**

A dishwashing machine must automatically dispense detergents and sanitizers. These machines must have a visual means to verify detergents and sanitizers are delivered or a visual or audible alarm to signal if the detergent and sanitizers are not delivered to the respective washing and sanitizing cycles. The machine must be large enough to accommodate the size and volume of equipment and utensils to be cleaned and sanitized.

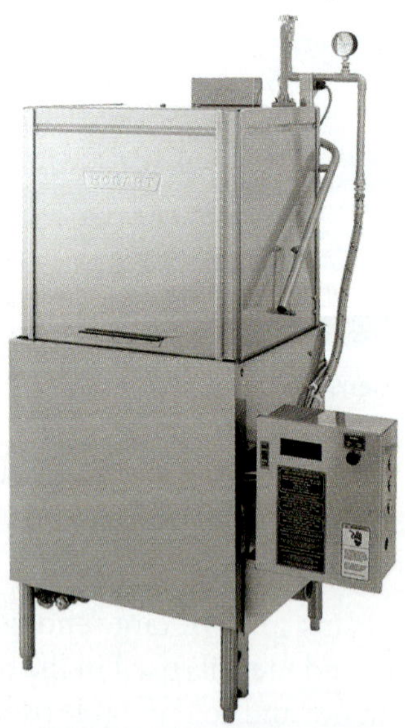

Single-tank Dishwashing Machine

(Courtesy of Hobart Corp.)

Low-temperature dishwashers are similar in design to the single-tank, stationary-rack dishwashing machine. However, they use chemicals to sanitize equipment and utensils. This allows lower water temperatures, which conserves energy.

When purchasing a dishwashing machine, you should also consider the cost of operation and maintenance. An adequate supply of very hot water is required for the final rinse in a high-temperature dishwashing machine. The high temperatures required for effective sanitization make it expensive and unsafe to maintain general-purpose hot water at these temperatures. Therefore, a separate booster heater is needed to raise the temperature of the sanitizing rinse water to the proper temperature. The booster heater must be properly sized, installed, and operated to deliver water at the volume, flow pressure, and temperature required for the operation.

Installation

Proper installation is required to assure equipment functions properly. The best design and construction will be worthless if electrical, gas, water, or drain connections are inadequate or improperly installed. The dealer who sells you equipment may or may not be responsible for its installation. Arrangements for installation will usually be specified in your purchase agreement. Following installation, employees must be trained to operate equipment correctly and safely.

Maintenance and Replacement

The cost of care and upkeep on a piece of equipment may determine whether or not its purchase and use are justified. Successful maintenance of equipment requires definite plans to prolong its life and maintain its usefulness. Such plans place emphasis on a few simple procedures:

- Keep the equipment clean
- Follow the manufacturer's printed directions for care and operation
- Post the instruction card for a piece of equipment near it
- Stress careful operation and maintenance schedules
- Make needed repairs promptly.

Some valuable suggestions for the care of equipment are:

- Assign the care of a machine to a responsible person
- Check the cleanliness of machines daily
- Have repairs performed promptly and by a properly trained person.

Lubricants used on food equipment may directly or indirectly end up in food. Therefore, all lubricants used on food-contact surfaces, bearings and gears, and other components of equipment that are located so the lubricants may leak, drip or be forced into food or onto food-contact surfaces must be approved as food additives or generally recognized as safe. These lubricants should be used in the smallest amount needed to get the job done.

Lighting

Proper lighting in production and dishwashing areas:

- Shows when a surface is soiled and when it has been properly cleaned
- Increases productivity
- Improves workmanship

Glass and food don't mix.
Courtesy of Shat-R-Shield, Inc.

- Reduces eye fatigue and employee irritability
- Decreases accidents and waste due to employee error.

Food production and warewashing areas should be furnished with the proper amount of lighting and soft colors that reduce glare. Locate lights to eliminate shadows on work surfaces or glare and excessive brightness in the field of vision.

The amount of light required in a department or area depends on the kind of work performed there. The following light intensity levels are recommended in the *FDA Food Code*:

Recommended Light Intensity Levels

10 foot-candles at a distance of 30 inches above the floor

- Walk-in refrigeration units
- Dry food storage areas
- Other areas and rooms during periods of cleaning.

20 foot-candles at a distance of 30 inches above the floor

- Areas where fresh produce and packaged foods are offered for sale and consumption
- Areas used for handwashing
- Areas used for dishwashing
- Areas used for equipment and utensil storage
- Toilet rooms.

50 foot-candles at the work surface

- Where an employee is working with unpackaged, potentially hazardous food
- Where an employee is working with food, utensils, and equipment such as knives, slicers, grinders, or saws
- Where employee safety is an overriding concern.

The food regulations in some state and local jurisdictions may require higher illumination levels in retail food establishments, especially during periods of cleaning. Consult the food regulation in your jurisdiction to determine what lighting levels are required for your store.

Light bulbs must be shielded, coated or otherwise shatter-resistant when used in areas where there is exposed food; clean equipment, utensils, and linens; or unwrapped single-service and single-use utensils. This is to prevent glass fragments from getting into food and onto food-contact surfaces. Shielded or shatter-resistant bulbs are not required in areas where packaged foods are stored and will not be affected by broken glass falling onto them, and the packages are capable of being cleaned of debris from broken bulbs before being opened.

Plastic coated shatterproof bulbs
Courtesy of Shat-R-Shield, Inc.

Light bulbs over self-service buffets and salad bars must also be shielded or shatter-resistant. In addition, this lighting must not mislead customers about the quality of the food items on display.

Use shielded lighting over exposed foods.

Heating, Ventilation, and Air Conditioning (HVAC)

Air conditioning in retail food establishments means more than simply "cooling the air." It includes heating, humidity control, circulation, filtering, and cooling of the air. HVAC systems must filter, warm, humidify, and circulate the air in the winter and maintain comfortable air temperature in the summer.

The National Fire Protection Association recommends all hoods be equipped with a fire extinguishing system.

Ventilation in food production and warewashing areas is typically provided by mechanical exhaust hood systems. These systems keep rooms free of excessive heat, steam, condensation, vapors, obnoxious odors, smoke, and fumes. The standard ventilation system used in retail food establishments consists of a hood, fan, and intake and exhaust air ducts and vents. Ventilation hood systems must be designed and constructed to prevent grease or condensation from dripping onto food, equipment, utensils, linens, or single-service and single-use articles. The capacity of the ventilation system should be based on the quantity of vapor and hot air to be removed.

Ventilation hood system

Hoods are usually constructed of stainless steel or a comparable material that provides a durable, smooth, and easily cleanable surface. The hood should be equipped with filters or other grease-extracting equipment to prevent drippage onto food.

Filters and other grease-removal equipment must be:

- Easily removed for cleaning and replacement
- Designed to be cleaned in place.

Fires at retail food establishments have been caused by the combustion of grease that accumulates in filters and ducts. Cleaning and replacement of filters must be part of your routine maintenance program. Intake and exhaust air ducts must be cleaned so they do not become a source of contamination of dust, dirt, and other materials. If vented to the outside, ventilation systems must not create a public health nuisance or unlawful discharge.

Check It Out

Back to the Story… ***Shell eggs that have not been treated to destroy*** **Salmonella** ***bacteria must be stored and displayed in refrigerated equipment that maintains an ambient temperature of 45°F (7°C) or less. Vertical display refrigerator***

cases have a "safe load line." This line indicates the level below which foods must be stored to assure the food is held at the proper temperature. It is also important to store foods so the discharge or return air vents are not blocked. Some of the cartons of eggs should be removed to assure the eggs on display are stored at proper temperature.

Good planning is essential for maximum productivity and efficiency. The total allotted space needed and the arrangement of equipment in the space are the overall features you must consider when planning the layout of a retail food establishment. The design of your facility should promote the smooth flow of work, provide adequate facilities and space for performing work efficiently, and contribute to high sanitation standards.

✓ The basic needs of a retail food establishment will determine the type and size of equipment purchased. When purchasing equipment, look for items that:

- Will serve the current and future needs of the operation
- Can be properly cleaned and sanitized
- Do not require extraordinary maintenance and repair.

✓ The *FDA Food Code* and construction standards from the American Standards Institute (ANSI) require food equipment and utensils to be:

- Smooth
- Easily cleanable
- Easy to take apart
- Easy to reassemble
- Equipped with rounded corners and edges.

✓ Materials used as food-contact surfaces must be safe, durable, corrosion-resistant, nonabsorbent, and resistant to chipping, pitting and deterioration. These materials must not allow the transfer of harmful substances, colors, odors, or tastes to food.

✓ Always install equipment in accordance with local building, plumbing, electrical, health, and fire safety codes. The best possible design and construction of equipment is worthless if electrical, gas, water, or drain connections are inadequate or the equipment is poorly installed.

Quiz 5 (Multiple Choice)

Choose the BEST answer for each question.

1. Floor-mounted equipment on legs must be a minimum of ______ inches above the floor.

   a. 2.
   b. 4.
   c. 6.
   d. 8.

2. The cooling efficiency of a refrigerator is greatly reduced by:

   a. Spacing products evenly throughout the storage area.
   b. Placing products on shelves 6 inches above the floor.
   c. Loosely covering the foods to be cooled in the unit.
   d. Lining the shelves of the refrigerator with paper or aluminum foil.

3. Which of the following materials is not recommended for use as a food-contact surface in retail food establishments?

   a. Soft wood.
   b. Stainless steel.
   c. Plastic.
   d. Chrome-plated steel.

4. Good sanitation requires adequate lighting in food production areas in order to:

   a. Provide a comfortable work environment for employees.
   b. Show when a surface is soiled and when it has been properly cleaned.
   c. Decrease accidents and waste due to employee error.
   d. Reduce glare that causes eye fatigue.

5. Which of the following statements about ventilation is false?

   a. Ventilation is typically provided by means of a mechanical exhaust hood system.
   b. Ventilation hood systems must be designed and constructed to prevent grease or condensation from dripping onto food and food-contact surfaces.
   c. The hood should be equipped with filters or other grease-catching devices to prevent drippage into food.
   d. Intake and exhaust air ducts do not need cleaning if filters or other grease-catching devices are provided to collect grease and other condensation.

Answers to the multiple-choice questions are provided in Appendix A.

References/Suggested Readings

Food and Drug Administration and Conference for Food Protection (1997). *Food Establishment Plan Review Guide.* Washington, DC.

Food and Drug Administration (2001). 2001 *Food Code* - Chapter 4. U.S. Public Health Service. Washington, DC.

Food and Nutrition Service (1999). *A Guide for Purchasing Food Service Equipment.* U.S. Department of Agriculture. Washington, DC.

Longrèe, K., and G. Armbruster (1996). *Quantity Food Sanitation.* Wiley Interscience. New York, NY.

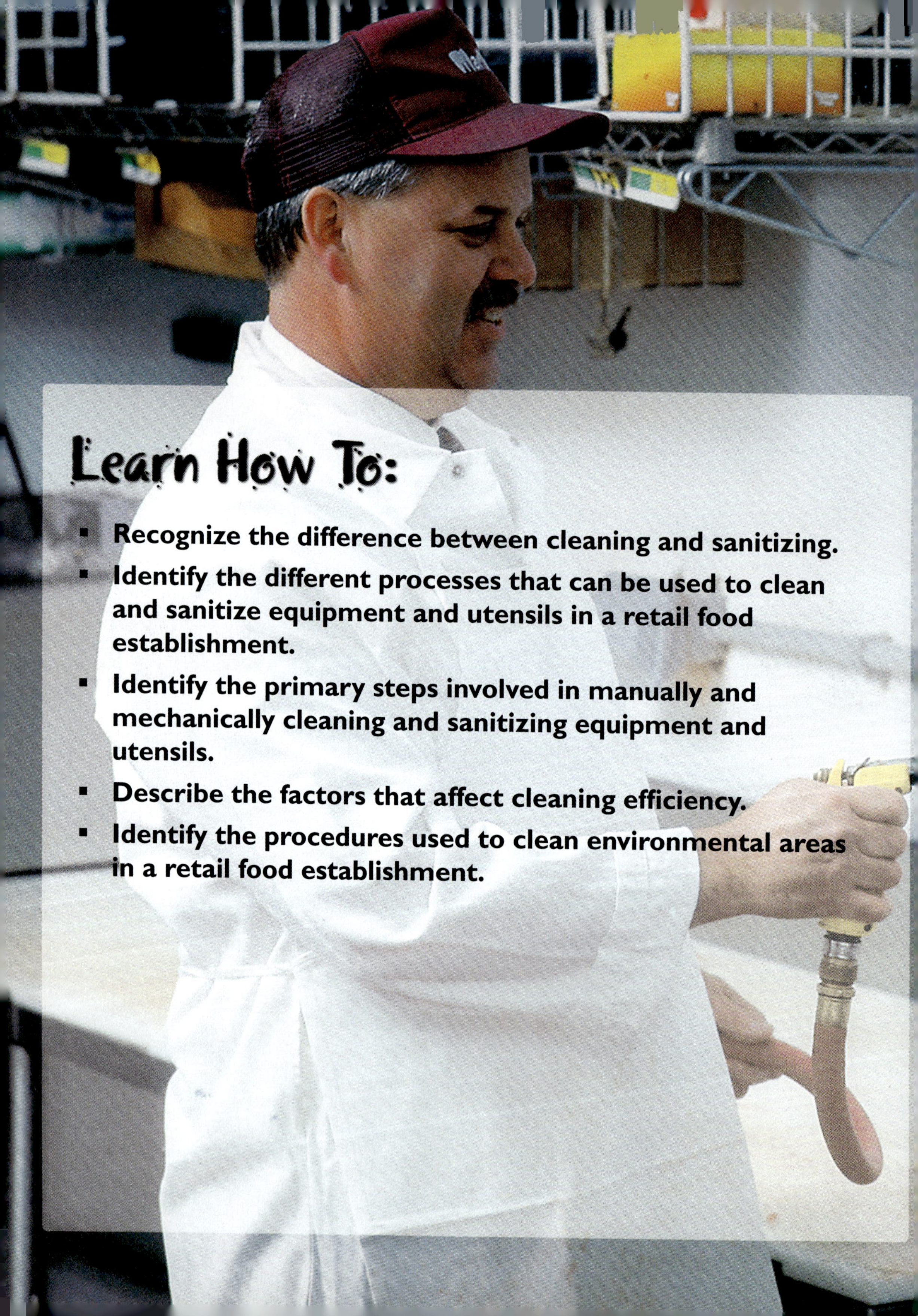

Learn How To:

- Recognize the difference between cleaning and sanitizing.
- Identify the different processes that can be used to clean and sanitize equipment and utensils in a retail food establishment.
- Identify the primary steps involved in manually and mechanically cleaning and sanitizing equipment and utensils.
- Describe the factors that affect cleaning efficiency.
- Identify the procedures used to clean environmental areas in a retail food establishment.

CHAPTER 6

Cleaning and Sanitizing Operations

Salmonellosis Outbreak at Cookout Celebration...

The meat department manager in a small grocery was in a hurry to get a special order of 12 salmon steaks for a customer. Shifts had just changed and he asked the nearest employee to prepare the order - which he did as quickly as possible. He did not know the knife and chopping block had just been used to split turkey breasts a few minutes earlier. The area was usually reserved for seafood only.

A week later the health department was notified of a foodborne disease outbreak of Salmonellosis. All ten victims had attended the same cookout and had eaten the steaks. An investigation uncovered the problems that led to this incident.

What went wrong? How could it have been avoided? See "Check it Out" at the end of this chapter to sort out the errors.

Principles of Cleaning and Sanitizing

Cleaning and sanitizing are important activities in all retail food establishments. Proper cleaning and sanitizing enhance the safety and quality of food and increase the life expectancy of equipment and facilities.

Cleaning and sanitizing are two distinct processes used for very different purposes. Cleaning is the physical removal of soil from surfaces of equipment and utensils. Most of the soil is food wastes and residues. The equipment and supplies used for cleaning are different from those used for sanitizing. Single-use items do not need to be cleaned. They must be discarded after use.

Sanitizing (sometimes called sanitization) is the treatment of a clean surface to reduce the number of disease-causing microorganisms to safe levels.

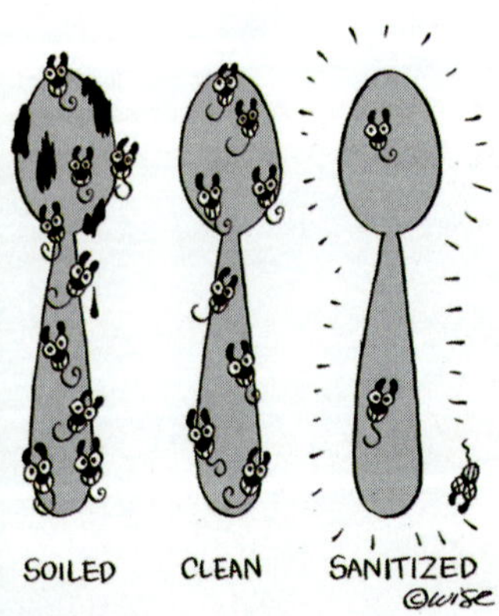

Effective Cleaning Consists of Four Separate Events:

1. A detergent or other type of cleaner is brought into contact with the soil.
2. The soil is loosened from the surface being cleaned.
3. The loosened soil is dispersed in the wash water.
4. The dispersed soil is rinsed away along with the detergent to prevent it from being redeposited onto the clean surface.

Removal of Food Particles

Scrape and flush food particles from the surfaces of equipment and utensils before the items are placed in a cleaning solution. Use warm water when pre-flushing and pre-scraping equipment and utensils. Avoid using very hot water or steam because they tend to "bake" food particles onto the surface of equipment and utensils and that makes cleaning more difficult.

Pre-flushing makes washing easier.

Application of Cleaning Agents

A cleaning agent is a chemical compound formulated to remove soil and dirt. It is important to choose the cleaning agent that is right for the particular job you have. There are many methods of applying cleaning agents to surfaces of equipment.

Cleaning agents typically include an acid or alkaline detergent and may include degreasers, abrasive materials or a sanitizer.

Soaking

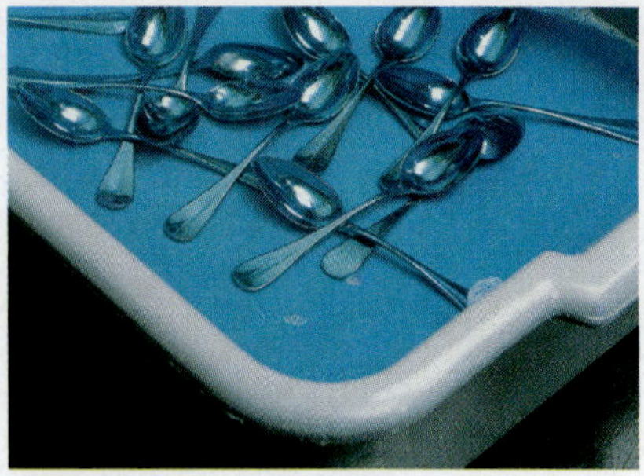

Soaking

Small equipment, equipment parts, and utensils may be immersed in cleaning solutions in a sink. By soaking equipment or utensils for a few minutes before scrubbing, you will increase the effectiveness of manual and mechanical warewashing.

Spray Methods

Spray cleaning

Cleaning solutions can be sprayed on equipment surfaces by using either fixed or portable spray units that use hot water or steam. These methods are used extensively in meat departments of retail food establishments.

Clean-in-place Systems

The clean-in-place method is an automated cleaning system generally used in conjunction with permanent-welded pipeline systems such as milk pasteurization equipment. The strength and velocity of the cleaning solution moving through the pipes are chiefly responsible for removing soil in clean-in-place (CIP) operations.

Abrasive Cleaning

Abrasive cleaners, in the form of powders and pastes, are used to remove soil firmly attached to a surface. Always rinse these cleaners completely and avoid scratching the surface of equipment and utensils. Abrasive type cleaners are not recommended for use on stainless steel surfaces. Never use metal or abrasive scouring pads on food-contact surfaces because small metal pieces from the pads may promote corrosion or may be picked up in food to become a physical hazard.

Rinsing

Immediately after cleaning, thoroughly rinse all equipment surfaces with hot, potable water to remove the cleaning solution. This very important rinse step is necessary because the product or detergent used for washing can interfere with the effectiveness of the sanitizer.

Factors Affecting Cleaning Efficiency

Factors	Impact on Cleaning	Description
Type of soil to be removed	Type of soil determines cleaning agents and process used to remove soil	Soil consists of: • Food deposits (proteins, carbohydrates, fats and oils) • Mineral deposits (salts) • Microorganisms (bacteria, viruses, yeasts, and molds) • Dirt and debris.
Water quality	Affects soap efficiency	• Must be potable (safe to drink) water • Cleaning agents must be compatible with the characteristics of your water supply.
Detergent or cleaner to be used	Cleaning agent or solvent—dissolves dirt and soil	• See cleaning chart on page 171 for more detailed information.
Water temperature	Increased water temperature helps decrease the strength of the bonds that hold soil to the surface	• Hot enough to remove soil but not to bake it on • Heat-stable detergents work best when the water temperature is 130°F (54°C) - 160°F (71°C).

(Continued)

Factors Affecting Cleaning Efficiency (Continued)		
Factors	**Impact on Cleaning**	**Description**
Velocity or force	Removes soil and film from food-contact surfaces	• Scrubbing or force to move soil/film from food-contact surfaces • Less force is required when detergent is working correctly.
Amount of time detergent/ cleaner remains in contact with the surface	Reduces scrubbing necessary to remove soil	• Soaking items increases cleaning efficiency.
Concentration of cleanser	Using the recommended amount of detergent improves cleaning power	Using too much detergent is a waste of money and may not improve cleaning.

Detergents and Cleaners to Be Used

The origin of the word is from the Latin, *detergeo*, meaning "to wipe away." Water acts as a detergent when soils are readily soluble. However, we can improve the cleansing action of water by adding soap, alkaline detergents, acid detergents, degreasers, abrasive cleaners, detergent sanitizers, or other cleaning agents to it. The water supply serving an establishment must be safe to drink (potable). Potable water is free from harmful microorganisms, chemicals, and other substances that can cause disease.

If there is an extended interruption in the water supply to a retail food establishment, the facility should cease operations or sell only prepackaged or pre-prepared food using single-use utensils until water service is restored.

Advantages and Disadvantages of Different Types of Detergents and Cleaners

Cleaning Agent	Advantages	Disadvantages
Soaps	• Effective for hand washing in soft water • Limited applications as cleaners	• Form precipitates and films in hard water • Not compatible with some sanitizers • Loses cleaning power in hard water.
Alkaline detergents	• Good general-purpose cleaners • Dissolves proteins and other organic material • Good buffers and enhance detergency	• Strong alkalis are corrosive and can harm metals, equipment surfaces, and skin.
Acid detergents	• Frequently used to remove food, mineral deposits, and hard water deposits from the surfaces of equipment and utensils	• Strong acids are corrosive to metals and irritating to skin.
Degreasers	• Remove grease and oily soils from hard surfaces • May be used for pre-treatment	• Can be irritating to skin and can leave a residue.
Abrasives	• When mixed with a detergent are useful for jobs that require scrubbing, scouring, or polishing	• Can scratch equipment surfaces • Abrasive particles may also contaminate food.

(Continued)

Advantages and Disadvantages of Different Types of Detergents and Cleaners (Continued)		
Cleaning Agent	**Advantages**	**Disadvantages**
Detergent sanitizers	• Effectively cleans and sanitizes a food-contact surface when applied to a food-contact surface two times—to clean the surface and to sanitize it • Product may be used at 2-bay sink where there is no distinct water rinse between the washing and sanitizing steps	• Can leave chemical residue if used at too high a level of concentration.

Cleaning Frequency

Under normal circumstances, food-contact surfaces and equipment used to prepare and serve potentially hazardous foods must be cleaned throughout the day to prevent the growth of microorganisms on those surfaces. Some guidelines for cleaning food-contact surfaces include:

- Before each use with a different type of raw animal food such as beef, fish, lamb, pork, or poultry, except when the surface is in contact with a series of different raw animal foods each requiring a higher cooking temperature
- Each time there is a change from working with raw foods to working with ready-to-eat foods
- Between uses with raw fruits and vegetables and with potentially hazardous food
- Before using or storing a food temperature-measuring device
- At any time during the operation when contamination may have occurred.

The *FDA Food Code* requires food-contact surfaces of equipment and utensils used to prepare potentially hazardous foods to be cleaned at least every four hours when used at room temperature.

There are some exceptions to the four-hour cleaning rule. One exception is where equipment and utensils are used to prepare

Retail food establishments must maintain records of the cleaning frequency based on the ambient temperature of the refrigerated room or area.

potentially hazardous food in a refrigerated room or area. The following chart shows the *FDA Food Code* prescribed cleaning frequency for refrigerated areas.

Room Temperature and Cleaning Frequency

Room Temperature	Cleaning Frequency
41°F (5°C) or less	at least once every 24 hours
41°F (5°C) - 45°F (7°C)	at least once every 20 hours
45°F (7°C) - 50°F (10°C)	at least once every 16 hours
50°F (10°C) - 55°F (13°C)	at least once every 10 hours

(Source: *FDA Food Code*)

In all cases, equipment and utensils that contact potentially hazardous foods must be cleaned every 24 hours.

Iced tea dispensers, carbonated beverage dispenser nozzles, water dispensing units, ice makers and ice bins are examples of equipment that routinely come into contact with food that is not potentially hazardous. These types of equipment must be cleaned on a routine basis to prevent the development of slime, mold, or soil residues that may contribute to an accumulation of microorganisms. The *FDA Food Code* recommends surfaces of utensils and equipment contacting food that is not potentially hazardous be cleaned:

- At any time when contamination may have occurred.
- At least every 24 hours for iced tea dispensers and customers self-service utensils such as tongs, scoops, or ladles.
- Before restocking customers self-service equipment and utensils such as condiment dispensers and bulk food display containers.
- Whenever possible, follow the manufacturer's guidelines for regular cleaning and sanitizing of the food-contact surfaces of equipment and utensils. If the manufacturer does not provide cleaning instructions, the person in charge should develop a cleaning regimen that will effectively remove soil, mold and other contaminants from equipment and utensils.

Guidelines for Cleaning Food-contact Surfaces that Touch Potentially Hazardous Foods

Item and Condition	When to be Cleaned
• Storage containers for potentially hazardous food when properly maintained for hot- and cold-holding temperatures	• Clean when empty
• Containers in serving situations (salad bars, delis, cafeteria lines) which are periodically combined with same food refills at required temperatures	• At least once every 24 hours
• Temperature-measuring device maintained in contact with food (deli food or roast)	• At least once every 24 hours
• Refrigerated equipment used for storage of packaged or unpackaged food (reach-in refrigerator)	• Frequency necessary to preclude accumulation of soil residue
• In-use utensils stored in water (containers with water at 135°F (57°C)	• At least once every 24 hours.

Sanitizing Principles

Heat and chemicals are the two types of sanitizers most commonly used in retail food establishments. Sanitizing does not provide the same level of microbial destruction as sterilization, because some bacterial spores and a few highly resistant vegetative cells generally survive the sanitizing process.

***Sanitizers* destroy disease-causing microorganisms that may be present on equipment and utensils even after cleaning.**

In all instances, a food-contact surface must be thoroughly cleaned and rinsed to remove soil and detergent residues before it can be properly sanitized.

Heat Sanitizing

Heat sanitizing in manual warewashing operations requires cleaned equipment and utensils to be submersed in hot water maintained at 171°F (77°C) or above for at least 30 seconds. A properly calibrated thermometer is needed to routinely check the temperature of the sanitizing water. Employees must use dish baskets or racks to lower equipment and utensils into the sanitizing water. Manual warewashing operations rarely employ this type of sanitizing due to concerns about employee safety and the large amount of energy required to keep the water hot.

Heat has Several Advantages over Chemical Sanitizing Agents Because it:

- Can penetrate small cracks and crevices
- Is noncorrosive to metal surfaces
- Kills all types of microorganisms equally effectively
- Leaves no residue
- Is easily measurable.

Sanitizing with hot water is most commonly performed as part of mechanical warewashing operations. The *FDA Food Code* requires the hot water used for these purposes to be between 180°F (82°C) and 194°F (90°C) when it leaves the final rinse spray nozzles. This assures the temperature of the water will be at 171°F (77°C) or above when it reaches the surfaces of the equipment and utensils being sanitized. The only exceptions to these temperature requirements are single-tank, stationary-rack, and single-temperature machines, where the final rinse water temperature must be at least 165°F (74°C) as it leaves the spray nozzles.

Steam is an excellent agent for treating pieces of food equipment. The equipment shown here is an example of a low-pressure, high temperature steam/vapor cleaning system. When steam in a flow cabinet is used, it should be sufficient to achieve 171°F (77°C) for at least 15 minutes or 200°F (94°C) for at least five minutes.

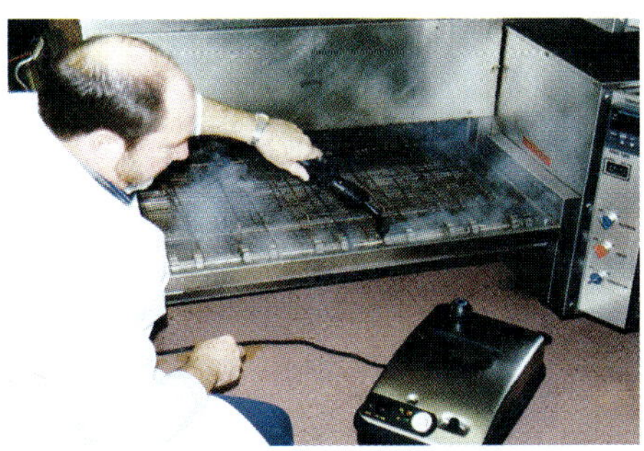

Low-pressure, high-steam cleaning system
Reprint with Permission of Food Safety Academy, Skokie, IL and Hi-Tech Cleaning Systems

Steam is sometimes produced in remote steam boilers. Most boiler systems will require the periodic addition of boiler water additives to prevent the buildup of scale and corrosion in the boiler system. When there is a chance steam produced in these boilers will come into direct or indirect contact with food, the chemicals added to boiler water must be approved for use as a food additive and labeled for that use as per Title 21 CFR Part 173.310—Boiler Water Additives.

When using heat for sanitizing, the temperature at the surface of the equipment and utensils must reach at least 160°F (71°C) to ensure proper destruction of disease-causing microorganisms. The temperature at the surface can be measured by using irreversible heat-sensitive labels or tapes attached to a clean dry dish and then run with a load of soiled dishes through a regular wash cycle. The labels change from white to black when the indicated temperature is reached.

Heat-sensitive label

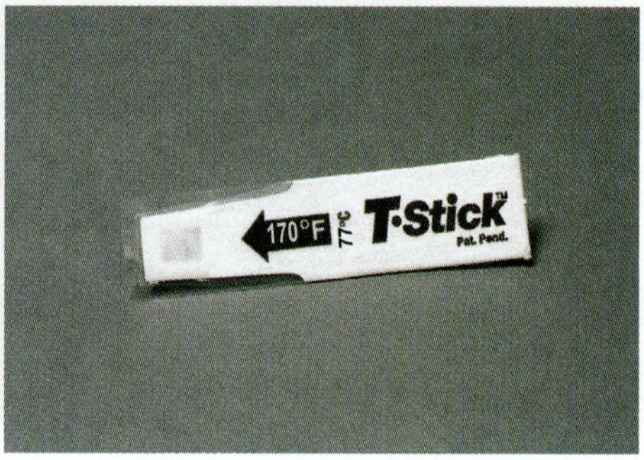

T-Stick

Measuring the sanitizing temperature in dishwashing machines

Another commonly used device for measuring the temperature of hot water sanitizers is the "maximum registering" or "holding" thermometer. This type thermometer will continue to hold the highest temperature measured until shaken down like a medical thermometer. An example of a maximum registering thermometer is presented is below. These instruments are becoming less popular because they contain mercury that can contaminate food and the general environment.

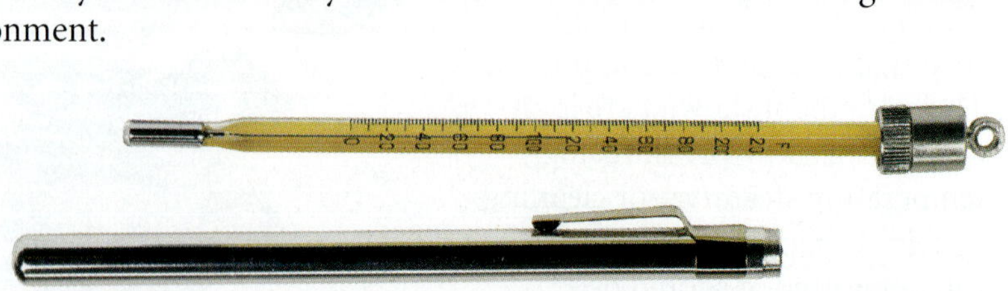

Maximum registering thermometers

Chemical Sanitizing

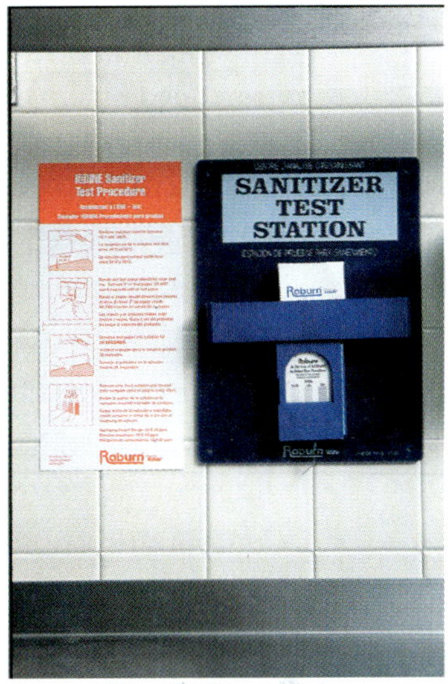

Chemical test strips

The chemical sanitizers most commonly used in retail food establishments are chlorine, iodine, and quaternary ammonium compounds (quats).

There are two ways to sanitize surfaces using chemical compounds:

- Immerse a piece of equipment or a utensil into a sanitizing solution at the prescribed concentration
- Swab, brush, or pressure spray the sanitizing solution directly onto the surface of the equipment and utensils.

A chemical test kit or test strips must be available so warewashing personnel can routinely check the strength of the sanitizing

Requirements for Sanitizing Equipment and Utensils			
Minimum Concentrations of Sanitizer Solution as expressed in parts per million (ppm) or mg/L	**pH 10.0 or less & Minimum Water Temperature**	**pH 8.0 or less & Minimum Water Temperature**	**Contact Time**
Chlorine 25	120°F (49°C)	120°F (49°C)	Greater than or equal to 10 seconds
Chlorine 50	100°F (38°C)	75°F (24°C)	Greater than or equal to 7 seconds
Chlorine 100	55°F (13°C)	55°F (13°C)	Greater than or equal to 10 seconds
Iodine Greater than or equal to 12.5 and Less than or equal to 25	pH less than or equal to 5.0 or per label and water temperature at 75°F (24°C) or above		Greater than or equal to 30 seconds
Quaternary Ammonium per label	slightly alkaline pH is preferred, water hardness must be 500 ppm or less or per manufacturer's label, and water temperature at 75°F (24°C) or above		
Hot Water Sanitize 3-compartment sink with integral heating device.	water temperature must be maintained at 171°F (77°C) or above and equipment and utensils completely immersed in rack or basket		
Steam	water temperature must be maintained at 200°F (94°C) or above		For at least 15 minutes

solution. The sanitizing solution should be replaced with a fresh solution whenever it becomes contaminated or if the concentration falls below the minimum level recommended by the manufacturer of the sanitizer.

Factors that Affect the Action of Chemical Sanitizers

The effectiveness of chemical sanitizers is affected by many different factors. Following are some of the most important factors to consider.

- Contact of sanitizer - in order for the chemical to react with and destroy microorganisms, it must achieve close contact.
- Selectivity of sanitizer - some sanitizers are nonselective and destroy a wide variety of microorganisms. Others exhibit a certain degree of selectivity.
- Concentration of sanitizer - in general, increasing the concentration of a chemical sanitizer proportionately increases its rate of microbial destruction. But there are limitations as the increased activity only extends to a certain maximum concentration, and then levels off. More is not always better, and high concentrations of sanitizers can be toxic and wasteful. Always follow the manufacturer's label use instructions to assure peak effectiveness of chemical sanitizers.
- Temperature of solution - all the common sanitizers increase in activity as the solution temperature increases. The standard range of water temperatures for chemical sanitizing solutions is between 75°F (24°C) and 120°F (49°C). Water temperatures as low as 55°F (13°C) can be used with chlorine under special circumstances, and temperatures above 120°F (49°C) should be avoided when using chlorine and iodine. At high temperatures, the potency of these sanitizers is lost by its evaporation into the atmosphere.
- pH or acidity of solution - water hardness can affect the pH of water that exerts a significant influence on most sanitizers. Most soaps and detergents are alkaline with a pH between 10 and 12. That is why soap and detergent must be rinsed off the surfaces of equipment and utensils before they are sanitized.
- Time of exposure - allow sufficient time for chemical reactions to destroy the microorganism. The amount of exposure time depends on the preceding factors as well as the size of the microbial populations and their susceptibility to the sanitizer.

Chlorine

Chlorine is a chemical component of hypochlorites. These compounds are commonly used as chemical sanitizers in retail food establishments. Hypochlorites are available as powders and liquids. The germicidal effectiveness of chlorine-based sanitizers depends, in part, on water temperature and the pH of the sanitizing solution.

Iodine

The iodine-containing sanitizers commonly used in retail food establishments are called iodophors. Iodophors are effective against a wide range of bacteria, small viruses and fungi. They are especially effective for killing disease-causing microorganisms found on human hands. Iodophors kill more quickly than either chlorine or the quaternary ammonium compounds. They function best in water that is acidic and at temperatures between 75° (24° C) and 120°F (49°C). Iodophors must be applied at 12.5 parts per million (ppm) when immersion sanitizing and at 25 parts per million in swab and spray applications.

Quaternary Ammonium Compounds

Quaternary ammonium compounds (quats) are ammonia salts used as chemical sanitizers in retail food establishments. Quats are effective sanitizers, but they do not destroy the wide variety of disease-causing microorganisms that chlorine and iodophor sanitizers do. Quats are noncorrosive and have no taste or odor when used in the proper dilution. Quats are more heat stable and can be used in hotter water than either hypochlorites or iodophors. At concentrations above 200 ppm, quats can leave a residue on the surface of an item. This is undesirable for food-contact surfaces. Always follow the manufacturer's directions for proper use and concentration of quats.

A summary of the advantages and disadvantages of the chemical sanitizers used most frequently in retail food establishments is shown on the next page.

Quats are not recommended at concentrations of 200 ppm or above for immersion sanitizing of food-contact surfaces.

Advantages and Disadvantages of Selected Chemical Sanitizers

SANITIZER	ADVANTAGES	DISADVANTAGES
Chlorine compounds	• Kills many types of microbes • Good for most sanitizing applications • Deodorizes and sanitizes • Nontoxic to humans when used at recommended concentrations • Colorless and nonstaining • Easy to handle • Economical to use	• Corrosive to equipment • Can irritate human skin and hands.
Iodophors	• Less corrosive to equipment • Less irritating to skin • Good for killing microbes on hands	• Moderate cost • Can stain equipment.
Quats	• Stable at high temperature • Stable for a longer contact time • Good for in-place sanitizers • Noncorrosive • No taste or odor	• Very expensive • Hard water can reduce effectiveness • Destroy a narrow range of microorganisms that may limit their use in some retail food establishments.

Mechanical Warewashing

Mechanical warewashing is performed in a dishwashing machine that cleans and sanitizes multi-use equipment and utensils automatically. Dishwashing machines are designed to clean and sanitize equipment and utensils that have no electrical parts and will fit into the machine. When properly maintained and operated, mechanical warewashing is as effective at removing soil and disease-causing microorganisms as is manual warewashing.

Single-tank dishwashing machine

Mechanical Warewashing Process

Mechanical warewashing uses an eight-step process to clean and sanitize equipment and utensils.

Mechanical Warewashing Process

1. Pre-scrape and pre-flush soiled equipment and pre-soak utensils to remove visible soil.
2. Rack equipment and utensils so wash and rinse waters will spray evenly on all surfaces and the equipment will freely drain.
3. Wash equipment and utensils in a detergent solution.
4. Rinse equipment and utensils in clean water at a temperature consistent with the type of dishwashing machine being used.
5. Sanitize equipment and utensils in a fresh, hot water sanitizing rinse between 180°F (82°C) and 194°F (90°C) for at least 30 seconds, except for a single-tank, stationary-rack, or single-temperature machine, where the final rinse may not be less than 165°F (74°C) for at least 30 seconds. The recommended final rinse temperature for low-temperature chemical sanitizing dishwashing machines is 120°F (49°C) or less.

   The flow pressure of the hot water sanitizing rinse must

be between 15 and 25 pounds per square inch (psi). The reduced water pressure is required in order to assure the sanitizing water covers the entire surface of the items being sanitized.

6. Drain properly and air-dry equipment and utensils. Do not towel dry food-contact surfaces.
7. Store clean and sanitary items in a clean, dry area where they are protected from contamination.
8. Clean and maintain the machine to keep it in proper working condition.

The temperature of the wash water (and power rinse if applicable) is extremely important in providing effective sanitizing of equipment and utensils. Water temperatures, water pressure, and conveyor speed or cycle time must be in accordance with the dish machine "data plate" and manufacturer's instructions.

Provide sufficient counter or table space for the accumulation of soiled equipment and utensils and keep them separate from items that are clean and sanitary.

WAREWASHING: MECHANICAL & MANUAL		Minimum Wash Temperature	Minimum Sanitizing Temperature
SPRAY TYPE WAREWASHERS : Single Tank, Hot Water Sanitize	Stationary rack, single temperature	165°F (74°C)	165°F (74°C)
	Stationary rack, dual temperature	150°F (66°C)	180°F (82°C)
	Conveyor, dual temperature	160°F (71°C)	
Multitank Hot Water Sanitize	Conveyor, multi temperature	150°F (66°C)	
Chemical Sanitize	Any warewashing machine	120°F (49°C)	sanitization levels as stated in the FDA Food Code or per labeled manufacturer's instructions on the container. *
3-Compartment Sink	Cleaning agent labeling may permit lower washing temperatures	110°F (43°C)	

* Chemical sanitization units installed after the adoption of the FDA's *2001 Food Code* must be equipped with an audible or visual "low level" sanitizer indicator to add more sanitizer.

Operators of dishwashing machines must be careful not to contaminate cleaned and sanitized equipment and utensils by touching them with soiled hands.

Cleaned and sanitized equipment and utensils must be properly drained and air-dried before storage. Soiled, multi-use cloth towels can recontaminate cleaned and sanitized equipment and utensils. The *FDA Food Code* prohibits cloth drying of equipment and utensils, with the exception that air-dried utensils can be "polished" with cloths that are clean and dry. If drying agents are used in conjunction with sanitization they must contain components that are Generally Recognized as Safe (GRAS) for use in food, GRAS for the intended purpose, or have been approved for use as a drying agent under the relevant provisions contained in Title 21 of the Code of Federal Regulations.

Cleaned and sanitized equipment and utensils must be stored in such a way as to protect them from contamination. Cabinets and carts used to store cleaned and sanitized equipment and utensils may not be located in locker rooms and toilet rooms and under sewer lines that are not shielded to catch drips, leaking water lines, open stairwells, or under other sources of contamination.

Proper storage of cleaned and sanitized equipment and utensils

Manual Warewashing

Manual warewashing typically uses a three-compartment sink to clean and sanitize equipment and utensils. The manual process begins with scraping and flushing food residues off the surface of equipment and utensils. When necessary, items may be pre-soaked to remove food residues and other soil.

After pre-flushing, the equipment and utensils are washed in the first compartment of the sink with warm water and a cleaning agent. Washing removes visible food particles and grease. Warewashing personnel should always use the correct amount of detergent based on the quantity of water used in the wash compartment of the sink. The temperature of the wash water must be maintained at not less than 110°F (43°C) unless the manufacturer of the cleaning agent specifies a different temperature.

Always make certain surfaces are properly cleaned and rinsed before they are sanitized.

Manual Warewashing			
	Compartment #1	**Compartment #2**	**Compartment #3**
Contains:	Detergent solution with hot water	Fresh warm potable water	Chemicals or hot water
Process:	Pre-scraped and pre-flushed equipment and utensils are washed	Soap and soil are rinsed off equipment and utensils	Sanitize the previously cleaned surfaces

The second compartment of the sink is used to rinse equipment and utensils in clean, warm water. Rinsing removes cleaning agents, soap film, remaining food particles, and abrasives that may interfere with the sanitizer agent. For best results, the rinse water must be kept clean and the temperature should be between 110°F (43°C) and 120°F (49°C).

Manual warewashing

The third compartment of the sink is used for sanitizing items with either hot water or chemical sanitizers.

When using hot water:

- Temperature must be maintained at 171°F (77°C) or above at all times
- Items must be immersed in the hot water for at least 30 seconds.

When using chemical sanitizers:

- The standard range of water temperatures for chemical sanitizing solutions is between 75°F (24°C) and 120°F (71°C).
- Under certain conditions chlorine sanitizers may be used with a water temperature as low as 55°F (13°C)
- Quats may be used with a water temperature above 120°F (71°C). Chlorine and iodophor sanitizers should not be used in water temperatures above 120°F (71°C) because the potency of these sanitizers is quickly lost at high water temperatures.

The manual warewashing sink must be equipped with sloped drain boards or dish tables of adequate size to store soiled items prior to washing, and clean items after they have been sanitized.

As with mechanical warewashing, items manually cleaned and sanitized must be air-dried before they are put into storage. If drying agents are used in conjunction with sanitization they must contain components that are GRAS for the intended purpose, or have been approved for use as a drying agent under the relevant provisions contained in Title 21 of the Code of Federal Regulations.

Hot and cold potable water must be supplied to each compartment of the sink, and the sink should be cleaned and sanitized prior to use.

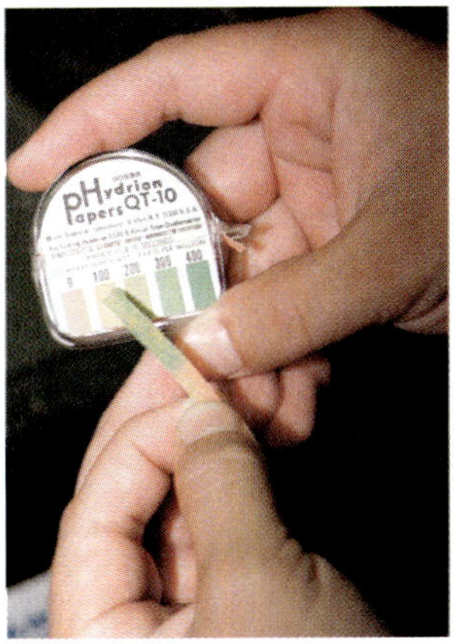

Measuring the concentration of sanitizers using test strips

The effectiveness of the manual warewashing process depends on the condition of the wash, rinse, and sanitizing solutions. Monitor wash and rinse solutions and replace them when they become soiled. When the sanitizer in the third compartment is depleted, it should be drained completely and replaced with a fresh solution at proper strength. The concentration of a chemical sanitizer solution must be tested periodically with a test kit or test strips to make sure it remains at strength. These strips provide a color comparison to indicate the strength of the sanitizer. Test kits or strips may be obtained from the manufacturer of your sanitizer.

Cleaning Fixed Equipment

Fixed equipment used in retail food establishments has food-contact surfaces but cannot be cleaned using either traditional mechanical or manual warewashing processes. Common examples of fixed equipment include band saws, floor-mounted mixers, slicers, grinders,

Cleaning fixed equipment

produce display cases and live tanks used to display live lobsters and molluscan shellfish. This equipment must be disassembled to expose food-contact surfaces to cleaning and sanitizing agents. The basic steps for cleaning fixed equipment are listed in the table below:

Steps for Cleaning Fixed Equipment

1. Disconnect the power supply to the equipment before taking it apart for cleaning.
2. Disassemble the equipment as necessary to allow the detergent solution to reach all food-contact surfaces.
3. Use a plastic scraper to clean equipment parts and remove food debris that has accumulated under and around the equipment. Scrape all debris into the trash.
4. Carry the parts that have been removed from the equipment to the manual warewashing sink where they will be washed, rinsed, and sanitized.
5. Wash the remaining parts of the equipment using a clean cloth, brush, or scouring pad and warm, soapy water. Clean from top to bottom.
6. Rinse thoroughly with fresh water and a clean cloth.
7. Swab or spray a chemical sanitizing solution, mixed to the manufacturer's recommendations, onto all food-contact surfaces.
8. Allow all parts to drain and air-dry.
9. Reassemble the equipment.
10. Resanitize any food-contact surface that might have been contaminated due to handling when the equipment was being reassembled.

Large equipment such as preparation tables and band saws can be cleaned by using a foam or spray method. In this process, detergents and degreasers, fresh water rinse, and a chemical sanitizer are applied using foam or spray guns. The hoses, feed lines, and nozzles that make up the foam or spray unit should be in good condition and attached properly.

Food-contact surfaces have to be cleaned and sanitized while non-food-contact surfaces require cleaning only.

The various display cases used in meat, seafood, dairy, bakery, produce and other departments in retail food establishments will have racks and other parts that can be removed for cleaning. Some equipment, such as produce and self-service bakery cases, may be equipped with accessory items such as misting heads and tongs or other handling devices. These accessories must also be cleaned and sanitized regularly to assure the safety and wholesomeness of products with which they are used.

Routine cleaning and maintenance of live seafood displays and holding tanks are necessary to preserve water quality and the quality of your live seafood. Some general guidelines that should be followed when working with this equipment are:

- Clean and sanitize the ultraviolet (UV) light unit used to kill pathogens in the water
- Clean the interior portion of the tank on a regular basis to keep it free of algae and slime buildup that may harbor pathogens
- If the tank is used for crab, lobster, etc., it must be cleaned and sanitized before oysters, clams, and other molluscan shellfish can be put into the tank.

The produce misting system in your establishment should be checked regularly to make sure it is clean and in good working order. Routine maintenance should include the following steps:

- If there is a separate water filter for the misting system, it should be checked and replaced as needed to assure an adequate supply of clean water
- Disassemble the nozzles and clean and sanitize the various parts to remove food debris and mineral residue
- Flush the tubing that supplies the mister heads with clean water
- Wipe the outside surfaces of the system (tubing, etc.) with an approved sanitizer to stop the growth of fungi, molds, and yeast.

Make sure your misting system is maintained regularly.

The bucket and brush method is often used to clean equipment that could be damaged by pressure spraying or immersion in the manual warewashing sink. This system uses one or more buckets for washing, rinsing, and sanitizing.

The clean-in-place (CIP) method is used for equipment that is designed to be cleaned and sanitized by circulating cleaning and sanitizing solutions through the equipment. Examples of equipment that would be cleaned using this method are soft-serve ice cream or yogurt machines. The basic steps in the clean-in-place method are listed in the following table.

Clean-in-place Cleaning Method

1. Empty food product and waste from the equipment.
2. Disconnect the power to the equipment.
3. Disassemble if parts are removable and cleanable in the manual warewashing sink.
4. Clean and sanitize the removable parts using the three-compartment warewashing sink process.
5. Clean and sanitize the main unit by circulating wash, rinse, and sanitizing solutions through it. Combination detergent sanitizers can also be used on equipment that is designed for CIP cleaning.
6. Reassemble parts removed from the main unit.
7. Resanitize by circulating a manufacturer approved sanitizing solution through the equipment.

Soft-serve ice cream machine
(Courtesy of SaniServ)

Many departments will use wiping cloths to clean up spills. These cloths should be stored in a container of sanitizing solution between uses in order to prevent microbial growth on the cloth. This is often referred to as an "in-place sanitizer." Any standard chemical sanitizer may be used for storing wiping cloths. However, quaternary ammonia compounds are preferred because they kill disease-causing microorganisms for a longer period of time.

- A clean cloth and fresh container of sanitizing solution at the proper strength must be used at the start of each day
- The cloth should be changed whenever it becomes heavily soiled and the sanitizer must be changed when it falls below the required concentration
- Wiping cloths may not be used for any other purpose except removing spills

- Cloths used to wipe spills from floors and non-food-contact surfaces must be kept separate from the cloths used to clean food-contact surfaces of equipment.

Store and use wiping cloths properly.

A retail food establishment should provide single-use and single-service articles for food handlers if it does not have proper facilities for cleaning and sanitizing multi-use kitchenware and tableware. These establishments must also provide single-service articles for use by consumers.

Materials used to make single-service and single-use articles must not permit the transfer of harmful substances or pass on colors, odors, or tastes to food. These materials must be safe and clean when used in retail food establishments.

Single-use items must be stored and dispensed in a way that will protect them from contamination from employees, customers, and the surrounding environment.

Cleaning Environmental Areas

A regular cleaning schedule of non-food-contact surfaces should be established and followed to maintain the facility in a clean and sanitary condition. The manager or supervisor should create a master-cleaning schedule that will list the following items:

- The specific equipment and facilities to be cleaned
- The processes and supplies needed to clean the equipment and facilities
- The prescribed time when the equipment and facilities should be cleaned
- The name of the employee who has been assigned to do the cleaning.

Cleaning environmental areas

Major cleaning should be done during periods when the least amount of food will be exposed to contamination and service will not be interrupted.

Ceilings and Walls

Ceilings should be checked regularly to make certain they are not contaminating food production areas. Ceilings, lights, fans, and covers can be cleaned using either a wet-or dry-cleaning technique. When wet-cleaning ceilings and fixtures, it is best to use a bucket method to keep water away from lights, fans, and other electrical devices. Walls may be cleaned using either the bucket or spray methods. Whenever possible, disconnect power before cleaning fixtures.

Floors

The cleanliness of floors depends on the ability of an employee to remove soil from the surface of the floor. Floors in customer shopping areas are usually cleaned with floor cleaning machines, and floors in food production areas can be cleaned using a spray system for washing and rinsing. Floors that will be damaged by spray cleaning can be cleaned using the bucket method.

All floors should be sloped to drains to help remove soil and the water used to clean floors.

If floor drains are not properly maintained, they can be a source of disease, vermin, and odors. All floor drains should be flushed, washed, rinsed, and sanitized.

The steps commonly used to clean floor drains are listed below:

1. Remove the grate or cover over the drain.
2. Clean out waste and other debris from the drain.
3. Use a sprayer or hose to flush the drain and grate or cover.
4. Pour in drain cleaner to break up grease and other waste in the drain.
5. Wash the drain using a brush or water pressure from the sprayer.
6. Rinse the drain with hot water.
7. Pour or spray sanitizer into the drain.

If not properly maintained, sewer pipes can become clogged and can cause sewage and other wastes to back up into and flood various areas of a retail food establishment. This can create both a health hazard and a nuisance. You must know what to do in the event a sewer backup occurs. Some of the steps you should follow are:

- Remove anything from the floor that might be damaged or contaminated by the wastewater.

- Remove the blockage in the pipe causing the backup. This may require the services of a professional plumber.
- Wash and sanitize all equipment, floors, walls, and any other objects that may have been contaminated by the waste that backed up into the establishment.

Equipment and Supplies Used for Cleaning

Some examples of equipment commonly used during cleaning are nylon brushes, cleaning cloths, scouring pads, squeegees, mops, buckets, spray bottles, hoses, and spray or foam guns. Commonly used cleaning supplies include hot water, cleaners, degreasers, and sanitizers.

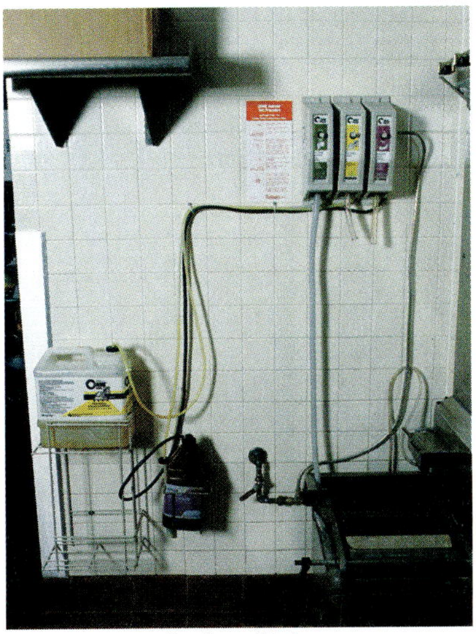

Janitor Sink

There should be a separate sink to fill and empty mop buckets, to rinse and clean mops and to clean brushes and sponges. A "janitor's" sink or floor drain should be provided to dispose of wastewater produced by cleaning activities.

Containers of poisonous or toxic materials and personal care items must bear a legible manufacturer's label. Working containers used for storing poisonous or toxic materials such as cleaners and sanitizers taken from bulk supplies must be clearly and individually identified with the common name of the material. A container previously used to store poisonous or toxic materials may not be used to store, transport, or dispense food.

The Occupational Safety and Health Administration (OSHA) requires employees to have the "Right-to-Know" about the chemicals to which they may be exposed on the job. Information about substances in the workplace is provided to employees by way of Material Safety Data Sheets (MSDS). Chemical manufacturers develop these and they must be maintained on file and accessible to employees in the retail food establishment.

Handwashing, food preparation, and warewashing sinks must never be used for cleaning mops and brushes.

Examples of information contained in an MSDS include:

- Ingredients
- Physical and chemical characteristics
- Fire, explosion, reactivity, and health hazard data
- How to handle chemicals safely
- How to use personal protective equipment and other devices to reduce risk
- Emergency procedures to use if required.

Failure to produce the required MSDS upon demand can result in a substantial fine from OSHA.

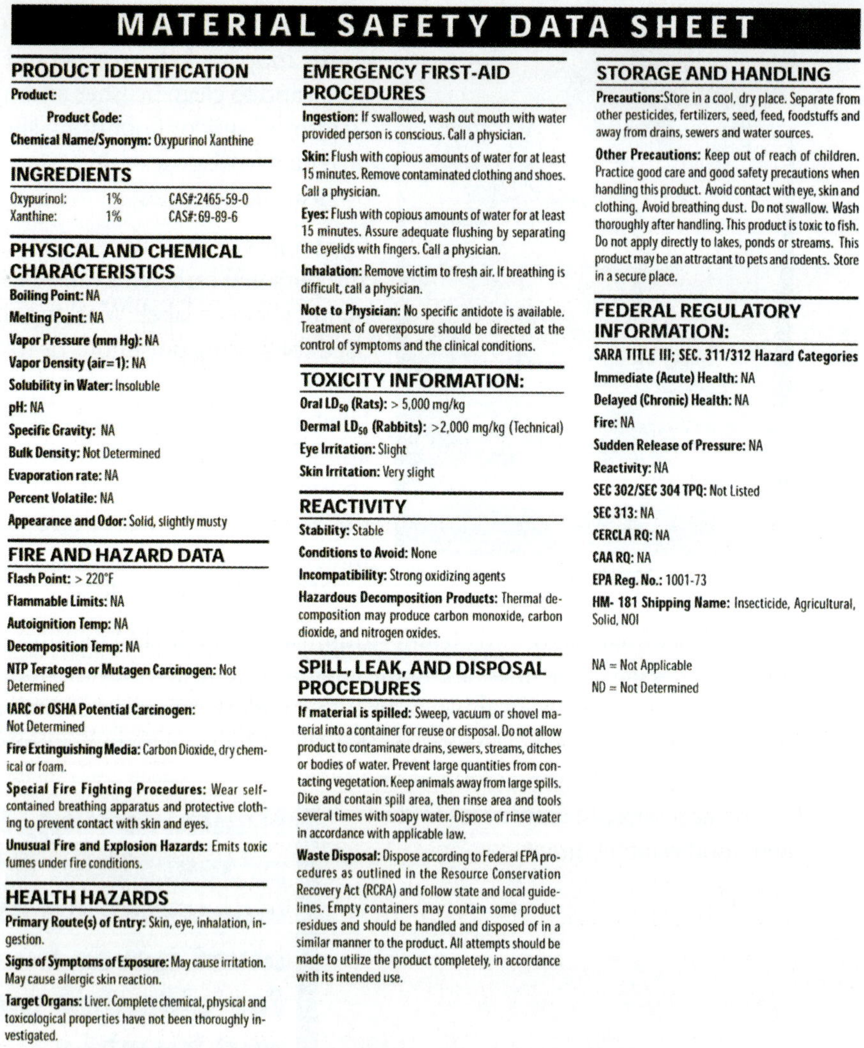

MATERIAL SAFETY DATA SHEET

PRODUCT IDENTIFICATION

Product:

Product Code:

Chemical Name/Synonym: Oxypurinol Xanthine

INGREDIENTS

Oxypurinol:	1%	CAS#:2465-59-0
Xanthine:	1%	CAS#:69-89-6

PHYSICAL AND CHEMICAL CHARACTERISTICS

Boiling Point: NA

Melting Point: NA

Vapor Pressure (mm Hg): NA

Vapor Density (air=1): NA

Solubility in Water: Insoluble

pH: NA

Specific Gravity: NA

Bulk Density: Not Determined

Evaporation rate: NA

Percent Volatile: NA

Appearance and Odor: Solid, slightly musty

FIRE AND HAZARD DATA

Flash Point: > 220°F

Flammable Limits: NA

Autoignition Temp: NA

Decomposition Temp: NA

NTP Teratogen or Mutagen Carcinogen: Not Determined

IARC or OSHA Potential Carcinogen: Not Determined

Fire Extinguishing Media: Carbon Dioxide, dry chemical or foam.

Special Fire Fighting Procedures: Wear self-contained breathing apparatus and protective clothing to prevent contact with skin and eyes.

Unusual Fire and Explosion Hazards: Emits toxic fumes under fire conditions.

HEALTH HAZARDS

Primary Route(s) of Entry: Skin, eye, inhalation, ingestion.

Signs of Symptoms of Exposure: May cause irritation. May cause allergic skin reaction.

Target Organs: Liver. Complete chemical, physical and toxicological properties have not been thoroughly investigated.

EMERGENCY FIRST-AID PROCEDURES

Ingestion: If swallowed, wash out mouth with water provided person is conscious. Call a physician.

Skin: Flush with copious amounts of water for at least 15 minutes. Remove contaminated clothing and shoes. Call a physician.

Eyes: Flush with copious amounts of water for at least 15 minutes. Assure adequate flushing by separating the eyelids with fingers. Call a physician.

Inhalation: Remove victim to fresh air. If breathing is difficult, call a physician.

Note to Physician: No specific antidote is available. Treatment of overexposure should be directed at the control of symptoms and the clinical conditions.

TOXICITY INFORMATION:

Oral LD_{50} (Rats): > 5,000 mg/kg

Dermal LD_{50} (Rabbits): >2,000 mg/kg (Technical)

Eye Irritation: Slight

Skin Irritation: Very slight

REACTIVITY

Stability: Stable

Conditions to Avoid: None

Incompatibility: Strong oxidizing agents

Hazardous Decomposition Products: Thermal decomposition may produce carbon monoxide, carbon dioxide, and nitrogen oxides.

SPILL, LEAK, AND DISPOSAL PROCEDURES

If material is spilled: Sweep, vacuum or shovel material into a container for reuse or disposal. Do not allow product to contaminate drains, sewers, streams, ditches or bodies of water. Prevent large quantities from contacting vegetation. Keep animals away from large spills. Dike and contain spill area, then rinse area and tools several times with soapy water. Dispose of rinse water in accordance with applicable law.

Waste Disposal: Dispose according to Federal EPA procedures as outlined in the Resource Conservation Recovery Act (RCRA) and follow state and local guidelines. Empty containers may contain some product residues and should be handled and disposed of in a similar manner to the product. All attempts should be made to utilize the product completely, in accordance with its intended use.

STORAGE AND HANDLING

Precautions: Store in a cool, dry place. Separate from other pesticides, fertilizers, seed, feed, foodstuffs and away from drains, sewers and water sources.

Other Precautions: Keep out of reach of children. Practice good care and good safety precautions when handling this product. Avoid contact with eye, skin and clothing. Avoid breathing dust. Do not swallow. Wash thoroughly after handling. This product is toxic to fish. Do not apply directly to lakes, ponds or streams. This product may be an attractant to pets and rodents. Store in a secure place.

FEDERAL REGULATORY INFORMATION:

SARA TITLE III; SEC. 311/312 Hazard Categories

Immediate (Acute) Health: NA

Delayed (Chronic) Health: NA

Fire: NA

Sudden Release of Pressure: NA

Reactivity: NA

SEC 302/SEC 304 TPQ: Not Listed

SEC 313: NA

CERCLA RQ: NA

CAA RQ: NA

EPA Reg. No.: 1001-73

HM- 181 Shipping Name: Insecticide, Agricultural, Solid, NOI

NA = Not Applicable

ND = Not Determined

Material Safety Data Sheet

Check It Out

Back to the Story... ***The manager was rushing an order at shift change. Because the employee was just beginning the shift, he did not know an area commonly reserved for seafood had been used for turkey breasts. If the previous shift had finished cleaning and sanitizing the equipment in the area, no problem would have occurred. Turkey, like most poultry, is commonly contaminated with*** Salmonella ***bacteria and should be handled in separate areas. The reason the outbreak occurred is because the salmon steaks were not cooked to a temperature high enough to destroy the*** Salmonella ***organisms. After interviewing the staff, the health inspector was able to determine the problem had happened because of cross contamination due to contaminated equipment and utensils.***

- ✓ A good sanitation program starts with a neat, clean, and properly maintained building. A well-maintained facility helps protect food products from contamination, makes proper stock rotation easier, prevents entrance of pests, reduces fire hazards, and contributes to the overall safety of the work environment. Proper cleaning and sanitizing also enhance the safety and quality of food and increase the life expectancy of equipment and facilities.
- ✓ A retail food establishment's sanitation program consists of the standards, policies, and procedures that assure proper cleaning and sanitation. In-store sanitation standards must incorporate the requirements of government regulations and manufacturers' guidelines.
- ✓ The effectiveness of a retail food establishment's sanitation program is directly related to the degree of commitment and concern demonstrated by top management. Managers must motivate their employees to follow prescribed sanitation practices and reward those employees who do so successfully.
- ✓ Proper education and training is required to assure proper sanitation practices are followed. Employees responsible for handling food, equipment, and cleanup of the building and grounds must be trained to accept sanitation as one of their key responsibilities. A clean, well-maintained facility will become a source of pride for employees.
- ✓ For a sanitation program to be effective, cleaning tasks must be scheduled and individual employees assigned to complete the tasks. Employees must be given instructions that include what they are to clean, how to clean it, and what tools and

supplies are required to effectively clean it. Properly trained employees must carry out cleaning activities, and they must be closely monitored by supervisors to identify and correct problems when they occur.

A good sanitation program is a preventive program that anticipates and eliminates potential hazards before they become serious problems. The value of an effective program is difficult to measure in terms of dollars. Nonetheless, the value of effective sanitation can be evident in the form of diminished problems and increased goodwill with customers and the local regulatory community.

Quiz 6 (Multiple Choice)

Please choose the BEST answer.

1. The process that results in the complete removal of soil and food residues from surfaces of equipment and utensils is called

   a. Cleaning.
   b. Sanitizing.
   c. Pre-scraping.
   d. Rinsing.

2. The treatment of a clean surface to reduce the number of disease-causing microorganisms to safe levels is called

   a. Cleaning
   b. Sanitizing
   c. Pre-scraping
   d. Rinsing

3. The Occupational Safety and Health Administration (OSHA) requires that all potentially harmful substances have a sheet on file that gives information for emergency treatment if an employee is injured with the product. This document is called the

   a. Material safety data sheet.
   b. Emergency instruction information form.
   c. Government format for harmful products.
   d. Material information and content form.

4. Which of the following sequences best describes the correct way to clean and sanitize equipment and utensils in a retail food establishment?
   a. Wash, rinse, sanititze and towel dry.
   b. Pre-flush, wash, sanitize, air-dry.
   c. Pre-flush, wash, rinse, sanitize, air-dry.
   d. Pre-flush, rinse, sanitize, air-dry.
5. The last step that should be taken when cleaning and sanitizing fixed equipment such as saws and slicer is:
   a. Rinse all surfaces with fresh, potable water.
   b. Allow all parts to drain and air-dry.
   c. Wash parts that have been disassembled in a 3-compartment sink or mechanical dishwasher.
   d. Resanitize any food-contact surface that may have been touched when reassembling the equipment.

Answers to the multiple-choice questions are provided in Appendix A.

References/Suggested Readings

Code of Federal Regulations. *2001. 21 CFR 173.310 - Boiler Water Additives*, U.S. Government Printing Office. Washington, D.C.

Code of Federal Regulations. *2001. 21 CFR 172 - Substances Generally Recognized as Safe*, U.S. Government Printing Office. Washington, D.C.

Code of Federal Regulations. *2001. 21 CFR 184 - Direct Food Substances Affirmed as Generally Recognized as Safe*, U.S. Government Printing Office. Washington, D.C.

Code of Federal Regulations. *2001. 21 CFR 186 - Indirect Food Substances Affirmed as Generally Recognized as Safe*, U.S. Government Printing Office. Washington, D.C.

Food and Drug Administration (2001). *2001 Food Code*. United States Public Health Service. Washington, DC.

Learn How To:

- Describe how the visual appearance of a retail food establishment can affect a customer's opinion about the cleanliness and sanitation of the operation.
- Identify the equipment and supplies that must be provided in a properly equipped restroom facility.
- Identify the main components of a properly equipped handwashing station.
- Explain how conveniently located handwashing facilities contribute to good personal hygiene.
- Identify the types of plumbing hazards that can have a negative effect on public health.
- Identify some types of pests and common signs of pest infestation that may be found in retail food establishments.
- Describe how integrated pest management is used to control pests in retail food establishments.
- Explain how proper disposal and storage of garbage and refuse help prevent contamination and pest problems in a retail food establishment.

CHAPTER 7

Environmental Sanitation and Maintenance

Bulk Food Storage and Display Bins Serve More Than Customers

The local food market is known for exotic ingredients, fresh produce, and hard-to-find items and enjoys a high volume of business. It uses a lot of self-service display bins for rice, beans, peppers, and other dried foods. Customers scoop up the amounts of these products they wish to purchase and place them into plastic bags. The foods are often spilled as they are put into the plastic bags, and the areas around the bins are difficult to keep clean. There are bins of candy and dried fruits displayed and dispensed in the same manner in the aisle next to the dried vegetables.

The company hired to provide cleaning and general maintenance of the establishment reported to the manager they had seen signs of rodent and insect infestation. Rub marks had been seen along the walls behind the display bins and rodent droppings had been seen near the bins and in the storeroom. Cockroach egg cases had been found near the raisin and candy containers.

The manager called the pest control operator (PCO) who was under contract with the facility and asked for an immediate visit. In the meantime, the local inspector arrived to conduct his annual survey of the facility. Evidence of pest infestation was noticed and the infraction was checked on the inspection form. Following the inspection the PCO and the market manager held a meeting. What measures should they take for the pest problem at the establishment?

Condition of the Establishment

A pleasant facility attracts customers.

The exterior of the facility must be free of litter and debris that could attract and harbor pests and ruin the appearance of the facility. Grass and weeds should be regularly mowed to eliminate harborage areas for insects, rodents, and other pests. Walking and driving surfaces should be constructed of concrete, asphalt, gravel, or similar materials to facilitate maintenance and control dust. These surfaces should also be properly graded to prevent rainwater from pooling and standing on parking lots and sidewalks. To attract customers, the exterior of the building should be clean, attractive and make a good impression.

Proper Water Supply and Sewage Disposal System

An adequate water supply and proper sewage disposal are vital to the sanitation of retail food establishments. The water source and ability to meet hot water generation needs should be sufficient to meet demands of the retail food establishment. Drinking water for retail food establishments must be obtained from an approved source. Most establishments will be connected to a public water system. However, when a private well or other nonpublic water system is used, it must be constructed, maintained, and operated according to the water quality requirements of the jurisdiction. Wells must be located and constructed in a manner that will protect them from sewage and other sources of contamination. Periodic sampling is required to monitor the safety of the water and to detect any change in water quality.

According to the *FDA Food Code*, water from a nonpublic water system must be sampled and tested at least annually and as required by state water quality regulations.

To render the water safe, a drinking water system must be flushed and disinfected before being placed into service. The *FDA Food Code* prescribes a drinking water system is flushed after construction, repair, or modification and after an emergency situation, such as flood, that may introduce contaminants to the system.

A reservoir used to supply water to a device such as a produce fogger must be maintained in accordance with the manufacturer's specifications. The reservoir must be cleaned at least once a week in accordance with manufacturer's specifications, or according to the following procedures, whichever is more stringent:

1. Drain and completely disassemble the water and aerosol contact parts.
2. Brush-clean the reservoir, aerosol tubing, and discharge nozzles with a suitable detergent solution.
3. Flush the complete system with water to remove the detergent solution and particle accumulation.
4. Rinse by immersing, spraying, or swabbing the reservoir, aerosol tubing, and discharge nozzles with at least 50 ppm hypochlorite solution.

The use of non-potable water sources in retail food establishments must be approved by the regulatory authority in the jurisdiction and may only be used for non-culinary purposes such as air handling systems, cooling systems, and fire protection.

Proper disposal of sewage greatly reduces the risk of fecal contamination of food and water. The *FDA Food Code* requires sewage from retail food establishments to be disposed through an approved facility that is:

- A public sewage treatment plant, or
- An individual sewage disposal system that is sized, constructed, maintained, and operated according to rules and regulations of the jurisdiction.

Condition of Building

The cleanliness and attractiveness customers view upon entering the building influence their overall shopping experience. Entrance doors should be self-closing to discourage flying insects. Clearly mark the doors for entrance and exit to prevent accidents between those entering and leaving the establishment.

The entrance area to a retail food establishment creates a lasting impression.

Floors, Walls, and Ceilings

Consider the specific needs of the different departments when selecting materials for floors, walls and ceilings. Some criteria that should be used for all departments are:

- Sanitation
- Safety

- Durability
- Comfort
- Cost.

Materials used for floors, walls, and ceilings in food preparation areas, store rooms (including dry storage areas and walk-in refrigerators), warewashing areas, and restrooms must be:

- Smooth
- Nonabsorbent
- Easy to clean
- Resistant to damage and deterioration.

Proper construction, repair, and cleaning of floors, walls, and ceilings are important elements of an effective sanitation program.

Floors

The *FDA Food Code* prohibits the use of carpeting in:

- Food preparation areas
- Walk-in refrigerators
- Warewashing areas
- Toilet room areas where handwashing lavatories, toilets, and urinals are located
- Refuse storage rooms or other areas subject to moisture.

Coving is a curved sealed edge between the floor and wall that eliminates sharp corners or gaps that would make cleaning difficult and ineffective.

Floors graded to drains are needed in retail food establishments where water-flush methods are used for cleaning. In addition, the floor and wall must be coved and sealed. Coving is a curved sealed edge between the floor and wall that eliminates sharp corners or gaps that would make cleaning difficult and ineffective. When cleaning methods other than water flushing are used for cleaning floors, the floor and wall juncture must be coved with a gap of no more than 1/32 inch (1 mm) between the floor and wall.

Use mats and other forms of anti-slip floor coverings where necessary to protect employees from slips and falls. These devices should also be impervious, nonabsorbent, and easy to clean.

Ant-slip mats help prevent falls.

Walls and Ceilings

Walls and ceilings in food production and warewashing areas must be made of a light colored material to enhance the artificial lighting in these areas. This will make soil and dirt easier to see and will help employees know when they have done an effective job cleaning a surface. Walls and wall coverings should be constructed of materials such as ceramic tile, stainless steel, or fiberglass when used in areas cleaned frequently.

Ceilings should be constructed of nonporous, easily cleanable materials. Studs, rafters, joists, or pipes must not be exposed in walk-in refrigeration units, food preparation and warewashing areas, and toilet rooms.

Light fixtures, ventilation system components, and other attachments to walls and ceilings must be easy to clean and maintained in good repair.

Restroom Sanitation

Toilet facilities near work areas promote good personal hygiene, reduce lost productivity, and permit closer supervision of employees. Toilet rooms in retail food establishments must be completely enclosed and provided with tight-fitting and self-closing doors.

Materials used in the construction of toilet rooms and toilet fixtures must be durable and easily cleanable. The floors, walls and fixtures in toilet areas must be clean and well maintained. Supply toilet tissue at each toilet. Provide easy-to-clean containers for waste materials and have at least one covered container in toilet rooms used by women.

Employee restrooms must be conveniently located and accessible to employees during all hours of operation.

Handwashing Facilities

Food employees must know when and how to wash their hands. Refer to Chapter 3 for details on the proper way to wash your hands. Conveniently located and properly equipped handwashing facilities are key factors in getting employees to wash their hands. Handwashing stations should be located in or near areas where food is prepared or handled and in warewashing areas. Handwashing stations must also be located in or adjacent to restrooms. The number of handwashing stations required and their installation are usually set by local health or plumbing codes.

Poor sanitation in toilet areas can spread disease.

Never store food or items associated with food handling in restroom areas.

A handwashing station must be equipped with:

- Hot and cold running water under pressure
- A supply of soap
- A way to dry hands without contaminating them.

A handwashing lavatory must be able to provide water at a temperature of at least 100°F (38°C) through a mixing valve or combination faucet. If a self-closing, slow-closing, or metering faucet is used, it must provide a flow of water for at least 15 seconds without the need to be reactivated.

Each handwashing station must be equipped with a dispenser containing liquid or powdered soap. The use of bar soap is frequently discouraged by regulatory agencies because bar soap can become contaminated with germs and soil.

Individual disposable towels and mechanical hot-air dryers are the preferred hand-drying devices. Most local health departments do not allow retractable cloth towel dispenser systems because there are too many possibilities for contamination. Common cloth towels, used multiple times by employees to dry their hands, are also prohibited.

Handwashing stations should be clean and well maintained and never be used for purposes other than hand washing.

Sinks used for preparing produce and warewashing must not be used for hand washing.

Plumbing Hazards in Retail Food Establishments

A properly designed and installed plumbing system is very important to food sanitation. The *FDA Food Code* includes many different components within the definition of a plumbing system. In particular, it identifies the

- Water supply and distribution pipes
- Plumbing fixtures and traps
- Soil, waste, and vent pipes
- Sanitary and storm sewers
- Building drains, including their respective connections and devices within the building and at the site.

Numerous outbreaks of gastroenteritis, dysentery, typhoid fever, and chemical poisonings have been traced to cross connections and other types of plumbing hazards in retail food establishments. The plumbing system in a retail food establishment must be maintained in good repair. When repairs to the plumbing system are required, they must be completed in accordance with applicable local ordinances and state regulations.

Cross Connections

A cross connection may be either direct or indirect. A direct cross connection occurs when a potable water system is directly connected to a drain, sewer, non-potable water supply, or other source of contamination. An indirect cross connection is where the source of contamination (sewage, chemicals, etc.) may be blown across, sucked into, or diverted into a safe water supply.

A *cross connection* is any physical link through which contaminants from drains, sewers, or waste pipes can enter a potable (safe to drink) water supply.

Backflow

Backflow occurs most frequently under two conditions:

- Backpressure where contamination is forced into a potable water system through a connection that has a higher pressure than the water system.
- Backsiphonage when there is reduced pressure or a vacuum formed in the water system. This might be caused by a water main break, the shutdown of a portion of the system for repairs, or heavy water use during a fire.

***Backflow* is the backward flow of contaminated water into a potable water supply.**

Backsiphonage

Methods and Devices to Prevent Backflow

The plumbing system in a retail food establishment must be designed, constructed, and installed according to the plumbing code in the local jurisdiction. A properly designed and installed plumbing system will keep food, equipment, and utensils from becoming contaminated with disease-causing microorganisms found in sewage and other pollutants.

Devices to Prevent Cross Connections and Backflow

Device	Key Features
Air gap 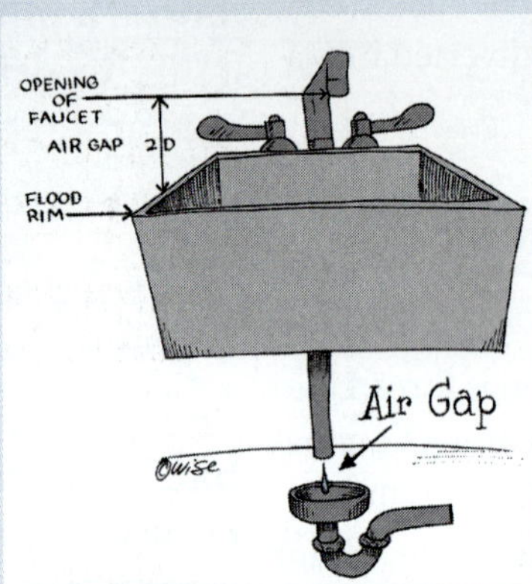 	• The most dependable backflow prevention device • Vertical air space separates potable and non-potable systems • Distance must be at least two times the diameter of the supply pipe (2D), but never less than 1 inch (25 mm).
Atmospheric vacuum breaker 	• Atmospheric vent used in combination with a check valve • Valve's air inlet closes when the potable water flows in the normal direction, but as water ceases to flow the air inlet opens, thus interrupting the possible backsiphonage effect.

(Continued)

Devices to Prevent Cross Connections and Backflow (Continued)

Device	Key Features
Pressure type vacuum breaker 	• Designed for use under continuous supply pressure, but not effective under backpressure • Use only where non-pressure vacuum breakers cannot be used.
Double check valve 	• Two internally loaded, independently operating check valves that may be used as protection for all direct connections through which foreign materials might enter a potable water system.
Reduced pressure principle backflow preventers 	• Used on all direct connections that may be subject to backpressure or backsiphonage • Used when there is a possibility of contamination by material that is a health hazard.

Backflow Prevention Devices on Carbonators

Carbonators on soft drink dispensers form carbonic acid by mixing carbon dioxide with water. The carbonic acid is then mixed with the syrups to produce the soft drinks. If the carbon dioxide backs up into a copper water line, the carbonic acid will dissolve some of the copper. The water containing the dissolved copper will then be used in dispensing soft drinks, and the first few customers receiving the drinks are likely to suffer the symptoms of copper poisoning. An air gap or a vented backflow prevention device meeting American Society of Sanitary Engineering (ASSE) Standard No. 1022 must be installed upstream from a carbonating device and downstream from any copper in the water supply line to reduce incidences of copper poisoning.

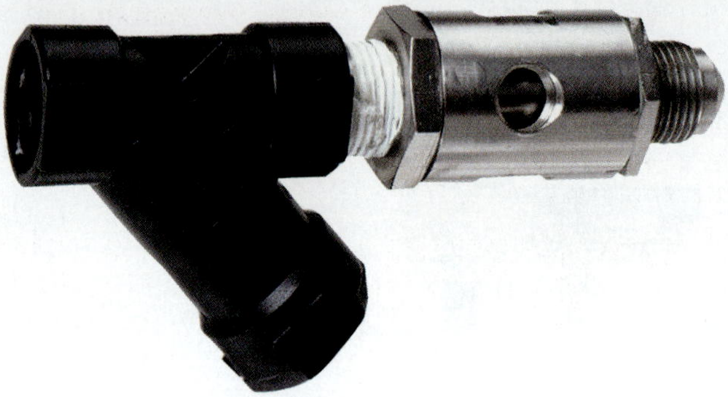

Backflow preventer on a carbonator

Grease Traps

Retail food establishments that produce large amounts of grease from a meat department and similar operations should be equipped with a grease trap. These devices remove liquid grease and fats after they have hardened and become separated from the wastewater. Grease traps are especially important when the retail food establishment is connected to a septic system or other type of onsite wastewater treatment and disposal system. A grease trap must be located for easily accessible cleaning.

Failure to properly clean and maintain grease traps can result in the harborage of pests and/or the failure of the facility's sewage system.

Grease trap spended from basement ceiling

Garbage and Refuse Sanitation

Proper storage and disposal of garbage and refuse are required at retail food establishments to protect food and equipment from contamination. Insects, rodents, and other pests are also less likely to be attracted to these establishments when garbage and refuse are properly managed. Effective waste management requires:

- Proper handling and short-term storage of the materials inside the operation
- Proper storage of the waste outside the building until it is picked up by a commercial refuse disposal company.

Garbage is the term applied to food wastes that cannot be recycled.

Refuse is trash, rubbish and other types of solid waste not disposed of through the sewage system.

Inside Storage

Waste containers must be provided in all departments and other areas in a retail food establishment where refuse is produced or discarded. Containers used to collect garbage and refuse must be:

- Durable
- Cleanable
- Insect and rodent-proof
- Leak-proof
- Nonabsorbent
- Covered with a tight-fitting lid when not in use.

Plastic bags and wet strength paper bags are frequently used to line waste containers. Do not place waste containers in locations where they might create a public health nuisance.

Waste storage container with plastic liner

Refuse storage rooms and containers must be cleaned as part of the establishment's routine cleaning program. Keeping the area clean is your best defense against pests. When cleaning this equipment, be careful not to contaminate food, equipment, utensils, linens, or single-service and single-use articles. Wastewater produced while cleaning the equipment and receptacles is considered to be sewage. It must be disposed of through an approved sanitary sewage system or other system constructed, maintained, and operated according to law.

Outside Storage

A retail food establishment should also have an outside storage area and enclosure to hold refuse, recyclables, and returnables awaiting pickup. An outdoor storage surface should be durable, cleanable, and maintained in good repair. Dumpsters and storage areas must be covered with tight-fitting lids, doors, or covers to discourage insects, rodents, and other types of pests.

Dirty equipment, containers, and waste facilities attract insects and rodents.

Outside garbage and refuse station

Refuse and garbage should be removed from the site as often as necessary to prevent objectionable odors and avoid conditions that attract or harbor insects and rodents.

Outdoor storage areas must be kept clean and free of litter. Suitable cleaning equipment and supplies must be available to clean the equipment and receptacles. Refuse storage equipment and receptacles must have drains and drain plugs must be in place.

Compactors and other equipment for refuse, recyclables, and returnables must be installed to minimize the accumulation of debris. Always make sure you clean under and around these units to prevent insect and rodent harborage.

Some retail food establishments may provide redeeming machines for recyclables or returnables. According to the *FDA Food Code*, a redeeming machine may be located in the packaged food storage area or consumer area of a retail food establishment if food, equipment, utensils and linens, and single-service and single-use articles are not subject to contamination from the machine and a public health nuisance is not created.

Trash bailer

Pest Control

Every retail food establishment should have a pest control program. The targets of this program are insects and rodents that can spread disease and damage food. These pests carry disease-causing microorganisms in and on their bodies and can transfer them to food and food-contact surfaces. Pests also destroy millions of dollars of food each year by eating it or by contaminating it with urine and feces.

The key element of a successful pest control program is prevention. However, no single measure will effectively prevent or control insects and rodents in retail food establishments. It takes a combination of three separate activities to keep pests in check. You must:

- Prevent entry of insects and rodents into the establishment
- Eliminate food, water, and places where insects and rodents can hide
- Implement an integrated pest management (IPM) program to control insect and rodent pests that enter the establishment.

Prevent pests from invading your establishment.

Insects

What insects lack in size, they more than make up for in numbers. Insects may spread diseases, contaminate food, destroy property, or be nuisances in retail food establishments. Insects need water, food, and a breeding place in order to survive. The best method of insect control is keeping them out of the establishment coupled with good sanitation and integrated pest management when needed.

Insects Common To Retail Food Establishments

Flies

Common Types:

- Houseflies
- Blowflies
- Fruit flies.

Common Problems:

- When a fly walks over filth, material sticks to its body and leg hairs, which contaminates food when a fly walks over it; the fly vomits on solid food to soften it before eating, spreading bacteria to food and food-contact surfaces
- Blowflies attracted by odors in food establishments
- Fruit flies attracted by decaying fruit.

Control Methods:

- Eliminate the insect's food supply; store food, garbage, and other wastes in fly-tight containers; regularly clean kitchen, dining, toilet, and waste storage facilities.
- Equip windows, doors, and loading and unloading areas with tight-fitting screens or air curtains.
- Insect electrocuting devices must be installed so dead insects and insect parts cannot fall on food and food-contact surfaces. Non-electrocuting systems, using glue traps and pheromone attractants, are allowed in areas of the establishment where electrocuting devices are not.
- Chemical insecticides may be applied by a professional pest control operator as a supplement to proper food-handling practices and a clean establishment.

Continued

Insects Common To Retail Food Establishments

Cockroaches	
Common Types:	• German cockroach 
Common Problems:	• Carry bacteria on their hairy legs and body as well as in their intestinal tract • Commonly hide in cracks and crevices under and behind equipment and facilities.
Control Methods:	• Maintain good housekeeping indoors and outside; eliminate hiding places by picking up unwanted materials; fill cracks and crevices in floors and walls and around equipment; doors and windows should be tight-fitting and protected by screening, air curtains, or other effective means • Check incoming food and supplies for signs of infestation such as egg cases and live roaches; store food in containers that are insect proof and have tight-fitting lids • Keep floors, tables, walls and equipment clean and free of food wastes • Residual insecticides and baits can be used when a serious infestation exists.

Continued

Insects Common To Retail Food Establishments	
Moths and beetles **Common Types:**	• Indian meal moth • Saw-toothed grain beetle • Flour weevil • Rice weevil 
Common Problems:	• Feed on corn, rice, wheat, flour, beans, sugar, meal, and cereals. • These insects create problems of wasted food and nuisance rather than disease.
Control Methods:	• Inspect incoming products for signs of infestation • Use FIFO system of stock rotation; store opened packages or bags of food in covered containers • Clean shelves and floors frequently • Keep dry food storage areas cool • Residual insecticides and pheromone traps are available to control these pests.

Rodents

Rodents are known to carry microorganisms that can cause a number of human diseases including salmonellosis, plague, and murine typhus. Rodents also consume and damage large quantities of foods each year. Rats typically carry their food back to their nest rather than eat it where it is found.

Domestic rodents in the United States include the *Norway rat*, the *roof rat*, and the *house mouse*. The Norway rat is also known as the brown rat, sewer rat, and wharf rat, and is the one most commonly found in the United States.

The *Norway rat* hides in burrows in the ground and around buildings and in sewers. Norway rats will eat almost any food but prefer garbage, meat, fish, and cereal. They stay close to food and water, and their range of travel is usually no more than 100 to 150 feet.

The *roof rat* generally harbors in the upper floors of buildings but is sometimes found in sewers. Roof rats prefer vegetables, fruits, cereal and grain for food. The range of travel for the roof rat is also about 100 to 150 feet.

House mouse
(Courtesy of Bell Laboratories, Inc.)

The *house mouse* is the smallest of the domestic rodents. It is found primarily in and around buildings, nesting in walls, cabinets, and stored goods. The house mouse is a nibbler, and it prefers cereal and grain. Its range of travel is 10 to 30 feet.

Signs of Rodent Infestation

It is unusual to see rats or mice during the daytime, since they are nocturnal. Therefore, it is necessary to look for signs of their activities. From rodent signs you can determine the type of rodent, whether it is a new or old problem, and whether there is a light or heavy infestation.

Droppings

The presence of rat or mouse feces is one of the best indications of an infestation. Fresh droppings are usually moist, soft and shiny, whereas old droppings become dry and hard. Norway rat droppings are the largest and have rounded ends. They look a lot like black jelly beans. Roof rat droppings are smaller and more regular in form. The droppings of the house mouse are very small and pointed at each end. They look something like dark grains of rice.

Rodent droppings

Runways and Burrows

Rats are very cautious and repeatedly use the same paths and trails. Outdoors in grass and weeds, you may see 2 to 3-inch wide paths worn down from repeated activity.

The Norway rat prefers to burrow for nesting and harborage. Burrows are found in earth banks, along walls, and under rubbish. Rat holes are about 3 inches in diameter whereas mouse holes are only about 1 inch in diameter. If a burrow is active, it will be free of cobwebs and dust. The presence of fresh food or freshly dug earth at the entrance of the burrow also indicates an active burrow.

Rub Marks

Rats prefer to stay close to walls where they can keep their highly sensitive whiskers in contact with the wall. As a rat runs along a wall, its body rubs against the wall or baseboard. The oil and filth from the rat's body are deposited on the wall and create a black mark called a "rub mark." Mice do not leave rub marks that are detectable, except when the infestation is especially heavy.

Gnawings

The incisor teeth of rats grow 4 to 6 inches a year. As a result, rats have to keep these teeth filed down in order to keep them short enough to use. Gnawings in wood are fresh if they are light colored and show well-defined teeth marks.

Tracks

Tracks may be observed along rat or mouse runs both indoors and outdoors. Look for tracks in dust in little-used rooms and in mud around puddles. Rat tracks may be 1 inch long.

Miscellaneous Signs

Rodent urine stains can be seen with an ultraviolet light (black light). Rats leave a different pattern than mice. Rat and mouse hairs may be found along walls, etc. When examined under a microscope, they can be distinguished from other animal hairs.

Rodent Control

The grounds around the retail food establishment should be free of litter, waste, refuse, uncut weeds, and grass. Unused equipment, boxes, crates, pallets and other materials should be neatly stored to eliminate places where pests might hide.

All entrances and loading and unloading areas should be equipped with self-closing doors and door flashings to prevent rodent entry into the establishment. Metal screens with holes no larger than ¼ inch should be installed over all floor drains to prevent entry.

Effective rodent control begins with a building and grounds that will not provide a source of food, shelter, and breeding areas.

Traps are useful around retail food establishments where rodenticides are not permitted or are hazardous. Live traps can be used for collecting live rats. Check traps at least once every 24 hours. Killer or snap traps can also be used as part of a rodent-control program. When using these types of traps, place them at right angles to the wall along rodent runways with the trigger side closest to the wall.

Bait box

Bait box

Glueboard

(Courtesy of J.T. Eaton & Company, Inc.)

Glueboards are shallow trays that have a very sticky surface. The mouse's feet stick to the board when it walks on it, and it is caught. Glueboards should be placed next to and running parallel with the wall.

Rodenticides are hazardous chemicals that can contaminate food and food-contact surfaces if not handled properly. Baits should be used outdoors to stop rodents at the outer boundaries of your property. Baits should be placed in a tamper proof, locked bait box that will prevent children and pets from being exposed to the toxic chemicals inside. Always make certain pesticides are stored in properly labeled containers, away from food in a secure place. Dispose of containers safely and know emergency measures for treating accidental poisoning.

The use of tracking powder pesticides is prohibited in retail food establishments. These types of pesticides can be dispersed throughout the establishment and directly or indirectly contaminate food, equipment, utensils, linens, and single-service/single-use articles. This contamination could adversely affect both the safety of the food and the general environment.

Integrated Pest Management (IPM)

Modern pest control operators use integrated pest management as the primary method to control pests in retail food establishments. Chemical pesticides are used only as a last resort and only in the amount needed to support the other control measures in the IPM program. If chemical pesticides are used, they must be approved for use as specified in the Code of Federal Regulations (40 CFR 152).

Integrated pest management (IPM) is a system that uses a combination of sanitation, mechanical, and chemical procedures to control pests.

The National Pest Management Association (NPMA) recommends a five-step program for IPM:

1. Inspection.
2. Identification.
3. Sanitation.
4. Application of two or more pest management procedures.
5. Evaluation of effectiveness through follow-up inspections.

There are many benefits produced by using an integrated pest management program. An IPM program is more efficient and cost effective than programs that rely exclusively on chemicals to control pests. IPM is also longer lasting and safer for you, your employees, and your customers.

Check It Out

Back to the Story... *The case presented at the beginning of the chapter illustrates how pests can pose a serious challenge for retail food establishments. Pests are known to contaminate*

food and food-contact surfaces, and they can carry disease-causing microorganisms harmful to humans. Pests are attracted to retail food establishments by the food, water and pleasant odors they find there. In order to keep pests under control, retail food establishments must prevent them from entering the facility, eliminate sources of food and shelter, and implement an effective sanitation program. Without these prerequisite activities in place, it would be impossible for the pest control operator to do his job properly.

- ✓ Surveys of customers show cleanliness is a top consideration when choosing a place to eat or shop for food. Customer satisfaction is highest in retail food establishments that are clean and bright and where quality food products are safely handled and displayed.
- ✓ Proper construction, repair, and cleaning of floors, walls, and ceilings are important parts of an effective sanitation program. Sanitation, safety, durability, comfort, and cost are the main criteria you will use when selecting materials for floors and walls. Surfaces of floors and walls should be resistant to damage and deterioration from the water, detergents, and repeated scrubbings used to keep them clean. Walls and ceilings should be light colored to show soil and enhance the artificial lighting used in food preparation, handling and display areas.
- ✓ Handwashing stations must be properly equipped and conveniently located to enable food handlers to wash their hands as necessary throughout the workday.
- ✓ Clean and suitably equipped toilet facilities must be provided for employees. These facilities must be kept clean and in good repair to prevent the spread of disease and promote good personal hygiene.
- ✓ A properly designed, constructed, and installed plumbing system is very important to food sanitation. Air gaps and mechanical vacuum breakers are used to protect the municipal water supply. Consult a professional plumber or your local plumbing code for details about the plumbing requirements in your jurisdiction.
- ✓ Proper storage and disposal of garbage and refuse are necessary to prevent contamination of food and equipment and avoid attracting insects, rodents, and other pests to a retail food establishment. Proper facilities and receptacles must be provided inside and outside the establishment to hold refuse, recyclables, and returnables that may accumulate. Refuse and garbage should be removed from retail food establishments frequently enough to minimize the development of objectionable odors and other conditions that attract or harbor insects, rodents, and other pests.

Quiz 7 (Multiple Choice)

Choose the BEST answer.

1. The most effective device for protecting the potable water system from contamination by backflow is a (an):
   a. Air gap.
   b. Double check valve.
   c. Hose bib.
   d. Vacuum breaker.
2. Coving is a (an):
   a. Curved sealed edge between the floor and wall that eliminates sharp corners to make cleaning easier.
   b. Anti-slip floor covering used to protect employees from slips and falls.
   c. Plastic material used to seal cracks and crevices under and around equipment in a food establishment.
   d. Device used to prevent backsiphonage.
3. The best way to encourage employees to wash their hands when needed is to:
   a. Provide a separate restroom for employees and customers.
   b. Provide properly equipped handwashing stations convenient to work areas.
   c. Provide hand sanitizers instead of handwashing lavatories in food-preparation areas.
   d. Put up a sign in the employee locker room reminding them of the importance of proper handwashing.
4. The primary responsibility of retail food establishment managers in pest control is to assure:
   a. Good sanitation that will eliminate food, water, and harborage areas.
   b. Pesticides are applied safely.
   c. The pest control operator they use employs integrated pest management.
   d. The parking area is kept free of litter.

5. Which of the following statements about garbage and refuse sanitation is false?
   a. Proper waste handling involves short-term storage of materials inside and outside the establishment.
   b. Improperly stored garbage and refuse can attract pests to a retail food establishment.
   c. Tight-fitting lids are not required on waste containers when they are lined with plastic garbage bags.
   d. Containers used for storing trash and garbage must be cleaned regularly.

Answers to the multiple-choice questions are provided in Appendix A.

Suggested Reading/References

Bennett, G, W., J. W. Owens, and R.M. Corrigan (1997) *Truman's Scientific Guide to Pest Control Operations* (5th. Edition). Advanstar Communications. Cleveland, OH.

Code of Federal Regulations. *2001. 40 CFR 152 - Pesticide Registration and Classification Procedures.* U.S. Government Printing Office. Washington, D.C.

Food and Drug Administration and Conference for Food Protection (1997). *Food Establishment Plan Review Guide.* Washington, DC.

Kopanic, R. J., B.W. Sheldon, and C.G. Wright (1994). *"Cockroaches as Vectors of Salmonella: Laboratory and Field Trials," Journal of Food Protection,* Vol. 57., No. 2, pp. 125-32. February, 1994.

Longrèe, K. and G. Armbruster (1996). *Quantity Food Sanitation.* Wiley. New York, NY.

Olsen, Alan R. (1998). *"Regulatory Action Criteria for Filth and Other Extraneous Materials-Review of Flies and Foodborne Enteric Diseases," Regulatory Toxicology and Pharmacology*, 28: 199-211.

Learn How To:

- Identify procedures for managing accidents.
- Discuss the importance of a safety audit.
- Identify public health rules and regulations that pertain to accidents and crisis management.
- Review a first-aid plan.
- Discuss the need for a fire exit plan, fire drill practice, and how to use a fire extinguisher.
- Inform staff of all terms used to alert for emergencies, i.e., "Code Adam."
- Identify the types and required locations of fire extinguishers.
- Identify the responsibility to comply with Occupational Safety and Health Administration (OSHA) rules.
- Review procedures for crisis management due to loss of utilities.

CHAPTER 8

Accident Prevention and Crisis Management

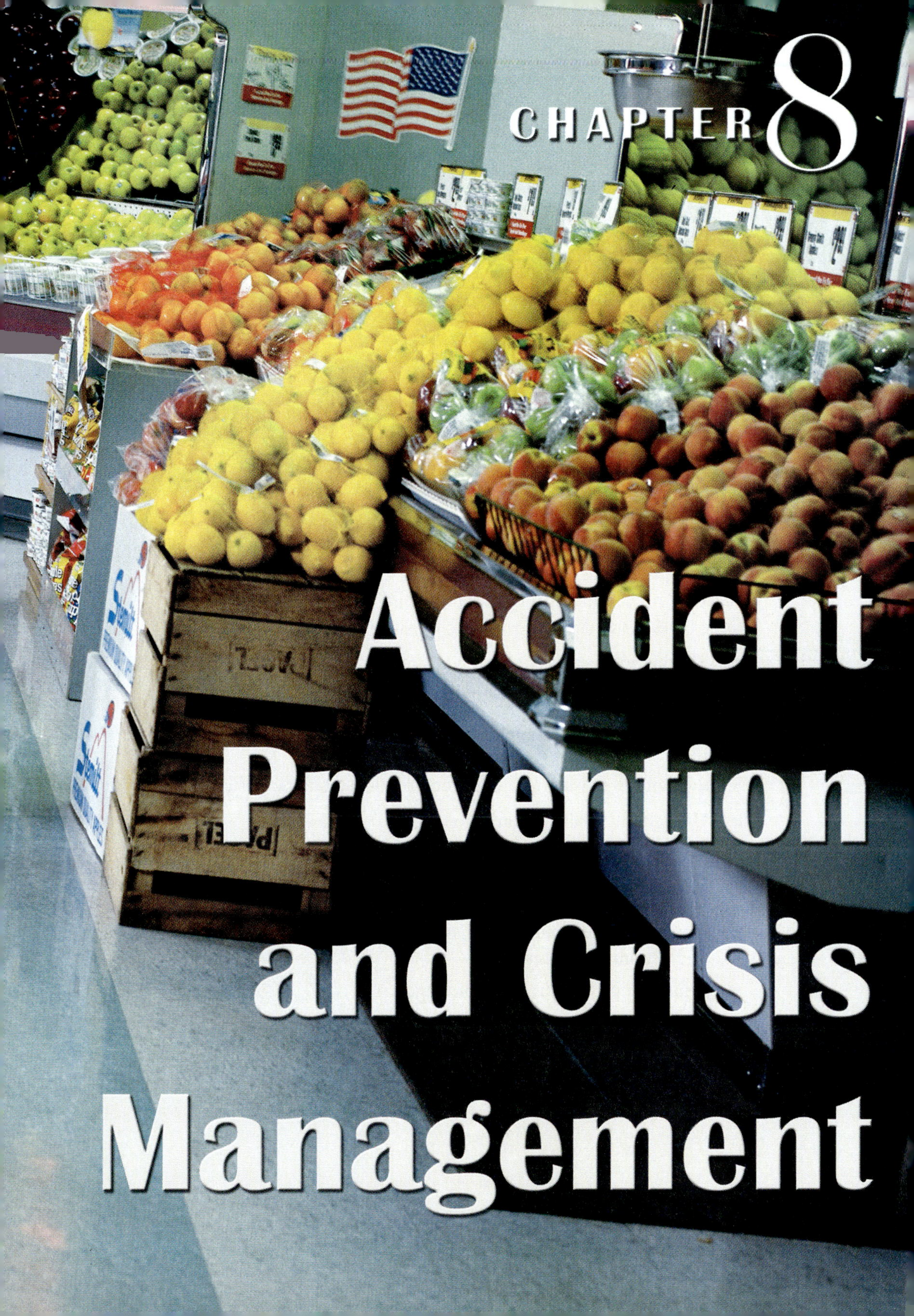

Awareness can Prevent Accidents

An employee who had received and stored a chemical order opened a box of liquid sanitizer in the storeroom. He used a case knife to open the box. One of the plastic bottles in the box was accidentally punctured allowing the liquid contents to slowly leak onto the floor unnoticed. While moving different products around the storeroom and into the sales area, the wheels of the hand truck tracked the sanitizer across the floor.

What are the potential problems that could occur as a result?

Safety in Retail Food Establishments

Retail food establishments contain many potential hazards that can cause accidents and injuries. Falls, cuts, scrapes, puncture wounds, and burns occur when food employees become careless or equipment is not in good repair.

Accidents, like foodborne illnesses, can be prevented.

Accident prevention programs typically involve three types of activities:

- Eliminating hazards in the environment
- Providing personal protective equipment for employees
- Training employees about potential hazards and how to prevent or avoid them.

Some examples of ways to prevent accidents in retail food establishments include:

- Using built-in safety guards to keep hands away from moving blades during grinding and slicing operations
- Wearing steel and nylon slash resistant gloves to prevent cuts when using slicers and deboning equipment
- Using hot pads to protect hands against burns
- Wearing work shoes with closed toes and rubber soles to shield the employee's feet from injury.

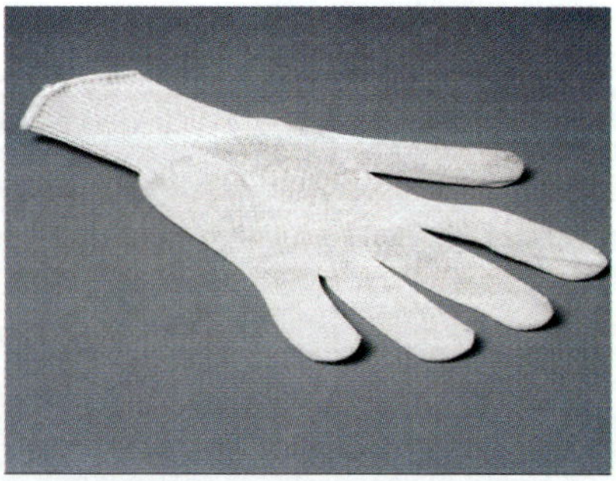

Slash resistent gloves

Employees must be taught how to prevent and avoid potential hazards that cause injury. Before they start to work, give employees instructions on specific procedures and safety measures related to their assigned tasks. They must know such things as:

- When and how to use ladders for climbing
- How to lift things properly
- How to use large heavy machinery such as a forklift, floor cleaning machine, compactor, etc.
- Why safety devices must be left in place on equipment
- How to dress for safety
- Minors are restricted from operating most power and mechanical equipment.

Dress for safety

Employees with wounds must cover the wound with an impermeable bandage and a disposable glove.

Each department needs a basic first-aid kit placed in a readily accessible area but away from all food, equipment, utensils, linens, and single-use articles. The kit should include an instruction manual describing basic first-aid treatments, sterile dressings, adhesive tape and bandages, burn cream, antiseptic ointment, and other basic first-aid supplies. Check expiration dates on supplies monthly and replace them as needed.

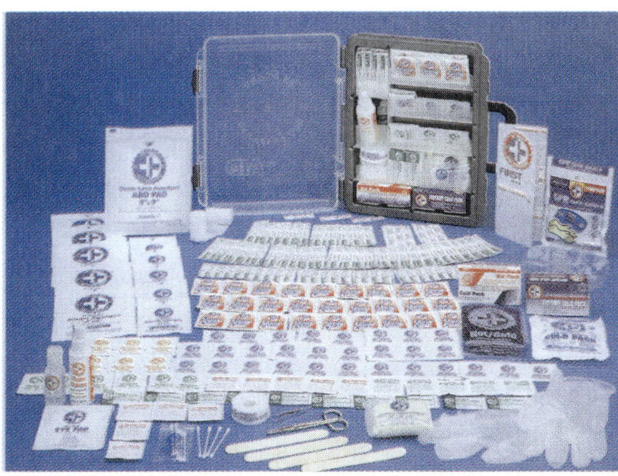

Keep a first-aid kit in an easily accessible area.

Post a list of important telephone numbers such as fire department, police, emergency medical system, and nearest hospital emergency room beside the telephone in a readily accessible area.

What to Do if an Accident Occurs

1. Stay calm. Information is needed to determine the seriousness of the injury. Begin by checking to see if the victim is responsive. Kneel and ask, "Are you OK? Would you like me to call for emergency medical services?"
2. Examine the injury and decide whether outside help is needed. If in doubt, call for help. Victims who are not breathing and do not have a heartbeat have a much greater chance of survival if they receive prompt medical care in a hospital or by trained paramedics.
3. Administer first aid according to the type of injury and your level of first-aid training.
4. Keep unnecessary personnel away from the victim.
5. Record the victim's name, the date and time of accident, type of injury or illness, any treatment given, and amount of time it takes for emergency assistance to arrive.

Common Types of Injuries

Common Types of Injuries		
Type of Injury	**Prevention**	**First Aid**
Cuts	• Keep knives and box cutters sharp and stored properly • Wear approved cut-resistant gloves • Don't leave knives on counters or submerged in dish water • Use proper tools for the task.	• Protect yourself from contact with body fluid • Apply pressure to cut with clean cloth or towel • Elevate the injury • Clean wound and apply antibiotic ointment and water-resistant bandage • A medical professional should treat severe cuts.

Continued

Common Types of Injuries (continued)		
Type of Injury	**Prevention**	**First Aid**
Falls	• Keep floors clean and free from debris • Use signs to alert others of hazards • Use anti-slip mats in high traffic areas.	• Keep victim still • Seek medical assistance if the victim has difficulty moving
Burns	• Provide hot pads for use • Don't use wet hot pads • Appropriately turn pot handles away from traffic areas • Vent steam from covered containers.	• Remove the source of the burn (including shutting off electricity if needed) • Soak in cool water • Cover the burn with a sterile bandage • Third-degree burns must be treated by a medical professional.

Protective gloves can prevent hand injuries.

Proper signage can prevent injuries.

Always use power equipment according to the manufacturer's directions. Leave guard devices in place—they protect you from injury. Most current equipment, such as slicers and grinders, has tension or safety switches. This means the equipment runs when the switch is manually held in place. Do not apply tape or any other type of bypass. The system is there to protect the user.

Always make sure slicers and other cutting equipment are unplugged before taking them apart for cleaning.

Follow posted warning signs.

Burns

Burns are classified as first degree (redness and pain), second degree (blisters, redness, and pain), or third degree (charring of skin layers, little or no pain). The extent of a burn is important. If a small area is burned, either first or second degree, first aid is usually all that is needed. A medical professional must check third-degree burns. Burn treatment includes the following:

Chemical burns require immediate treatment. Call for medical help and check the Material Safety Data Sheet (MSDS) for emergency interventions. Mishandling dishwashing compounds, cleaning solvents, scrubbing machines, and batteries are common causes of chemical burns.

- Remove the source of the burn. If the burn is caused by an electrical source, rescuers must first determine if the electricity is off before touching the victim.
- Soak the burned area in cool water to soothe minor burns and lower skin temperature.
- Over-the-counter pain medications may be used to help relieve pain and reduce inflammation and swelling.
- Cover the burn with a sterile bandage or a clean dry dressing. Do not puncture blisters.
- Keep the victim calm and quiet.

Poisoning

Report injuries to loss prevention manager or immediate supervisor.

Poisoning can occur when food contaminated by chemical substances is inhaled, consumed, or sold. Containers of poisonous or toxic materials and personal care items must bear a legible manufacturer's label. Working containers used for storing poisonous or toxic materials such as cleaners and sanitizers taken from bulk supplies must be clearly and individually identified with the common name of the material. A container previously used to store poisonous or toxic materials may not be used to store, transport, or dispense food. If there is any question a food is contaminated by chemicals, do not wait. Call for medical assistance immediately and begin internal recall crisis management. Contact the local health department if products have left the premises. Call poison control centers and follow instructions until help arrives.

Back Injuries

Back injuries cost the employer in lost productivity, medical care, and employee's compensation. A simple class and demonstration of good lifting and bending techniques is always a good investment.

The Food Marketing Institute has a series of videos on basic safety that are clear and easy to understand. The set includes a video on lifting and bending properly. Insurance companies and local health departments have resources to help educate employees on how to protect themselves from harm. Emergency medical service departments are often willing to come to the establishment to teach employees how to prevent injury.

Blood and Body Fluid-Borne Diseases

Any type of human blood must be handled with care. AIDS (Acquired Immunodeficiency Syndrome) and Hepatitis B are both blood-borne pathogens that can infect people.

Broken glass or other sharp objects should not be handled, even with gloved hands. Use a broom, brush, and dustpan to remove them to a designated container for special handling and disposal.

Body fluids and feces also carry harmful microbes and must be disposed of carefully. There are special procedures called universal precautions to protect employees from contact with potentially harmful microbes. The Food Marketing Institute materials or local health care facilities can supply instruction in special techniques required to protect humans.

Wear protective gloves.

Employees must wear protective gloves and outer garments and use cleaning and sanitizing procedures to remove blood or body fluids. Any food contaminated by blood or body fluids must be disposed of properly. A solution of 1 part chlorine bleach (or other commercial sanitizer) to 10 parts water should be used to disinfect any area contaminated with blood. Do not attempt to clean up a spill or assist another person without first obtaining gloves and protective gear for yourself.

When gloves are removed, pull the glove off turning it inside out. Drop it in the container with the contaminated material and then thoroughly scrub hands and forearms under warm running water. Any clothing that has been contaminated must be removed and washed. Check for any other skin contact and immediately flush and wash away splatters.

Federal law (OSHA 29 CFR 1910.i030) mandates employees should see a medical professional for care and follow-up. Always create an incident report to describe time, date, type of contact, and actions taken to protect the employee. The Food Marketing Institute has produced a video and learning materials about blood-borne pathogens and how to manage cleanup, disposal, and aftercare.

Cardiopulmonary Resuscitation (CPR) and First Aid for Choking

It is recommended members of your staff know how to administer first aid for choking or **Cardiopulmonary Resuscitation (CPR)**. Posting charts with instructions for these procedures and other lifesaving steps can be helpful in emergencies. These charts are available from the American Red Cross and the American Heart Association. It is desirable to have at least one person on each shift trained in emergency response procedures and first aid.

Knowing how to do the right thing at the right moment can save a life.

Lifesaving steps

(Source: American Red Cross - used with permission)

Employee Medications

Store employee medications away from food.

According to the *FDA Food Code*, only those medications necessary for the employee's health are allowed in a retail food establishment. This does not apply to medicines stored or displayed for retail sale. Employee medication must be clearly labeled and stored in an area away from food, equipment, utensils, linens, and single-use items like straws, eating utensils, and napkins. Medications stored in a refrigerator where food is stored must be kept in a clearly labeled and covered container located on the lowest shelf of the unit. Medications cannot be stored in display or storage cases, or walk-in units.

Self-inspection Safety Checks

Just as self-inspection programs help identify food safety problems in retail food establishments, safety audits can help prevent accidents. Conducting a safety audit can be one of the most cost effective tasks you can perform. Fire and accident prevention can result in considerable cost savings.

Fire Safety

Grease and electrical problems are common causes of fire in retail food establishments. Regular inspections of electrical equipment and prompt repair of faulty wiring or equipment can decrease this problem. Avoid overloading circuits and improper venting or grounding. Teach and monitor employees on the hazards of pulling electrical plugs out of sockets by the cord. They must grasp the plug head and then pull with dry hands.

In case of a fire emergency:

- Get customers and employees out of the building
- Call the fire department.

Keep hoods, ventilating fans, filters, ducts, and adjoining walls clean and clear of greasy buildup. Perform regular maintenance according to the manufacturer's directions.

Fire Extinguishers

Portable fire extinguishers must be operable, easy to find, and employed according to type of fire.

Type of Fire Extinguisher	Material Burning	Most Common Type of Extinguisher Used
Type A	wood, paper, cloth	Pressurized water or ABC
Type B	grease, gasoline, solvents	CO_2 or B or BC
Type C	motors, switches, electrical	CO_2 or C

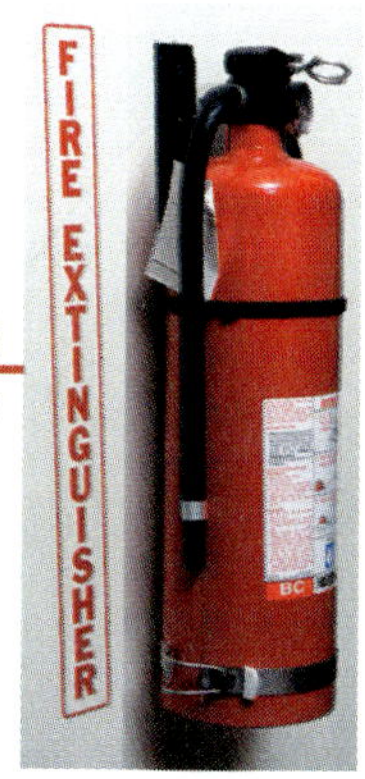

Types of fire extinguishers

Because employees' clothing can ignite when working around stoves or heat sources, a safety blanket should be stored in an area where it can easily be reached if needed. Use the blanket to smother the flames and then call for help.

Every retail food establishment needs a fire safety plan. Clearly mark routes of evacuation. Hold fire drills at periodic intervals to assure all employees know how and where to exit if the building is on fire. Battery-powered emergency lights are required in most jurisdictions. They automatically provide light when power is interrupted.

Never use a Type A fire extinguisher on a grease or electrical fire!

Fire safety inspectors check fire extinguishers to certify them as operable. Portable fire extinguishers should be checked or serviced at least annually and marked with tags that specify the next due date for service. Fines can be levied for out-of-date equipment.

If an extinguisher weighs less than 40 pounds, it can be mounted 5 feet from the ground. If it weighs more than 40 pounds, it may be stored no more than 3 feet off the ground.

Hood Systems

Some hood canopies may have a built-in extinguishing agent. These devices can effectively extinguish a fire. Employees need to know how they work. Hood systems should be installed and periodically serviced by fire prevention professionals, typically twice a year.

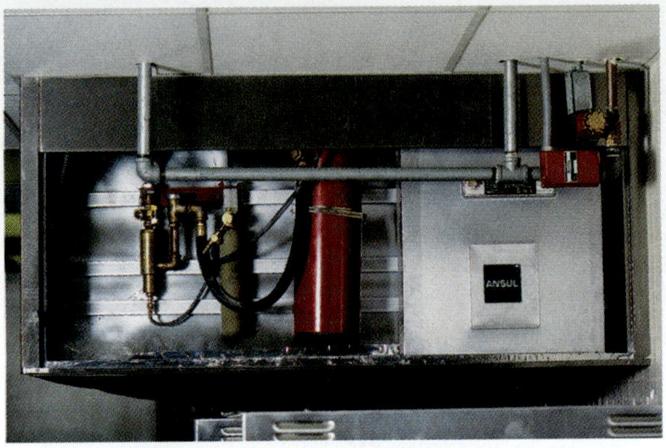

Fire extinguisher equipment

Sprinkler Systems

It is often required by law to install sprinkler systems in new or remodeled buildings. Supplies must be stored the regulated distance from sprinkler heads. If a sprinkler head is accidentally broken off, contact your sprinkler service immediately.

Operating a Forklift

Forklifts are common pieces of equipment in large retail food establishments. Any employee who operates a forklift must be certified through the Occupational Health and Safety Administration. Because this piece of machinery can easily become unbalanced and topple over or injure someone who may be standing under the lift forks, operators need to demonstrate knowledge of special handling techniques.

Always check operating manuals and local requirements before assigning personnel to operate a piece of equipment.

Crisis Management

A crisis can occur at any time. Besides natural disasters such as fire, flood, storms, or earthquakes, retail food establishments must deal with interference in normal operations that can literally cripple production. Other types of crises that can have a significant impact on retail food operations are power outages, interruption of water and sewer service, an outbreak of foodborne illness, media investigations and unexpected loss of personnel. How do you handle difficult situations? Develop a plan to handle the crisis with a problem-solving approach.

Evaluate the extent of the problem. Set priorities according to the resources that exist. Remain calm. Follow established rules and regulations as set up in codes. Identify any outside resources that can help solve the problem. Keep a record of actions and communications in case it should be needed. In all situations, be honest.

Most retail food establishments have a system of codes to alert employees about certain conditions.

Water Supply Emergency Procedures

On occasion a water utility will issue a boil water order when the bacteriological quality of the water it provides does not meet the requirements of the Safe Drinking Water Act. This commonly happens when a water main breaks or when the source of the drinking water is contaminated by runoff and other sources of pollution.

To continue operating under "boil water advisories/notices" or "interrupted water service" from all water supplies, retail food establishments must secure and use safe drinking water from an approved source. This can be bottled water or water hauled to the site in tank trucks. Bottled drinking water used or sold in a retail food establishment must be obtained from approved sources in accordance with the Code of Federal Regulations (21 CFR 129—*Processing and Bottling of Bottled Drinking Water*).

In emergencies, or as a temporary measure, water from contaminated or suspect sources can be disinfected by either chlorination or boiling. To chlorinate water, add six drops of liquid chlorine bleach to one gallon of water and mix. Check the label on the container to make sure the active ingredient, sodium hypochlorite, is available at a concentration of 5.25 percent. The important things to remember when chlorinating water are:

In emergencies, or as a temporary measure, water from contaminated or suspect sources can be disinfected by either chlorination or boiling.

- Wait thirty (30) minutes after adding the chlorine before using the water for drinking or cooking purposes.
- If the six drops of bleach does not give the water a slight taste of chlorine, double the amount of chlorine until a slight taste is present. The taste of chlorine will be evidence the water is safe to drink.

Another way to purify water is to boil it. Bring the water to a full boil for at least five minutes. Cool and aerate the boiled water by pouring it from one clean container to another. This will reduce the flat taste caused by boiling.

Alternative Procedures To Minimize Water Usage

- Commercially packaged ice may be substituted for ice made onsite
- Single-service items or disposable utensils may be substituted for reusable dishes and utensils
- Use prepared foods from approved sources in place of complex preparation on site
- Restrict menu choices or hours of operation
- Portable toilets may be utilized for sanitary purposes.

After the water emergency is officially lifted or water service resumes, these precautionary measures must be followed:

- Flush the building water lines and clean faucet screens and waterline strainers on mechanical dishwashing machines and similar equipment
- Flush and sanitize all water-using fixtures and appliances such as ice machines, beverage dispensers, hot water heaters, etc.
- Clean and sanitize all fixtures, sinks, and equipment connected to waterlines.

There must be water pressure before resuming operations in a retail food establishment, and the water should be sampled for bacteriological quality. The safety of water cannot be judged by color, odor, or taste. Don't hesitate to contact your local regulatory agency if you have any questions regarding appropriate operations at your establishment during water emergency orders.

Bioterrorism

On September 11, 2001, terrorism in the United States became a reality. For the past several years, the federal government and state agencies have developed plans to react to potential efforts of sabotage. Reacting in a rational manner is extremely important. The Food Marketing Institute (FMI) has developed guidelines to handle any suspected tampering or contamination problem in retail food establishments. There are several actions that can be used to manage these types of problems. These include the following:

- Monitor employees and allow only approved employees into production areas. The use of photo identification name tags helps to identify employees.
- Protect food preparation areas from anyone who is not assigned to that area.
- Review visitor policies and change rules to keep areas secure. Use sign-in and sign-out logs.
- Prohibit personal items like lunch containers, cases, purses, and other such items from processing areas. Provide storage for employee property in a separate room or locker.
- Report any unusual or suspicious activity to your supervisor or manager.
- If a suspicious problem is identified, call the Federal Bureau of Investigations (FBI) and FDA Office of Crime Investigation. These agencies are best equipped to handle emergency situations and will advise you of the steps to follow.
- Designate a spokesperson to deal with media or other inquiries. This method assures information is managed carefully and mixed messages are avoided.

Product tampering is also a food safety concern, and many retail food establishments have developed policies and procedures to deal with such problems. Managers should teach employees to be alert and to report any unusual activity immediately. It is always better to be overly cautious than to ignore a potential incident.

FMI has a *Pre-Emergency Planning and Disaster Recovery Manual* with phone numbers and instructions on what to do in cases of emergency. For more information about this manual, you may call the FMI Publications and Video Sales Department at (202) 220-0723.

FMI can also provide assistance and information for use when implementing product recalls. For help on recalls, call the food safety department at (202) 223-0661. FMI also has a document called "Measures to Improve Security in the Supermarket Industry." For more information on store security, call the above number.

Foodborne Illness Incident or Outbreak

If dealing with a suspected foodborne illness problem, cooperate with the customers and seek help from your local food regulatory agency. Work with the local regulatory agency to determine what may have caused the customer's illness. Establish the time the

If a foodborne illness is suspected, the first action is to remove all suspected product from sale.

suspect food was purchased and eaten. If a sample of the product is still available, preserve it for laboratory analysis. Ask the customer if medical attention has been sought and to save samples of any product that remains. Initiate your crisis management team members, which should include team members from your media affairs, food safety, loss prevention, and legal counsel.

Action Plan When Foodborne Illness Is Suspected

- Remove all affected food from the sales areas
- Isolate it in a separate area
- Label it so it does not become mixed with other foods
- Do not sell any more of the suspect product
- Listen to everything the customer says and look for clues that might explain the problem
- Do not try to diagnose the problem or belittle the customer's complaint
- Keep a record of the interview
- Assure the customer you will get back to him or her after investigating the problem
- Make sure you do contact the customer again after facts are gathered
- Do not admit your establishment is at fault—you have no proof of that until facts are checked.

Check the Product Source

If the reported item was prepared in the store, remove it from sale immediately. If two or more persons report the same problem, consult outside resources (local health department) on how to proceed. When preparing food samples for testing, it is wise to keep a sample in your freezer for cross validation of laboratory results. Public health officials, insurance agents, and, if needed, an attorney can help you manage the problem.

Formal Investigations

Remove suspect items from use. Health inspectors may want to conduct an inspection at the establishment. Employees may be required to have medical examinations. Closing a facility for a period of time for cleaning may be needed. Keep in mind whatever must be done should be accomplished as quickly as possible. Cooperation to protect public health is the end goal. In the final analysis, it does not matter where the error occurred. It must be corrected and prevented from happening again.

Whatever the crisis may be, use a calm, systematic approach and evaluate actions as much as possible. Crisis management is not an easy task. It is always useful to have some kind of protocol to guide you through the event. Put together a team or committee consisting of management and employees to regularly brainstorm worst-case scenarios. Plan disaster drills to keep employees informed and trained in crisis management skills. Designate a spokesperson responsible for talking to the media, should they become involved. Keep a notebook with instructions on how to proceed in emergency situations.

Check it Out

Back to the Story... ***The punctured bottle of sanitizer poses several problems. A wet floor can cause an employee to fall. Inventory accidentally set down on the floor would be contaminated. A person bending over to tie a shoe could contaminate hands that touch a wet shoestring. The sanitizer could be tracked into food preparation or sales areas. There are many outcomes of not knowing this hazard exists. Management of materials in a storeroom is very important.***

- ✓ Accident prevention programs are good investments. Use community resources to help build an effective plan. Insurance companies and vendors can supply training aids on accident prevention. Fire departments and emergency health-care providers can provide demonstrations and advice that prepares you for accidents or other problems linked to safety.
- ✓ Check the emergency lights to see if they are on as required.
- ✓ All equipment with on/off switches should be turned off until power is restored. Take care of all perishable items according to procedures. The Food Marketing Institute's *Pre-Emergency Planning and Disaster Recovery Manual* (1996) contains detailed instructions to manage all refrigerated, frozen, and dairy product materials during loss of power.
- ✓ Throwaway items must be disposed of according to health department rules to keep customers from obtaining discarded products that could cause harm.

Quiz 8 (Multiple Choice)

Select the BEST answer.

1. When slicing or removing bones from fresh meat, the food employee should:
   a. Always wear single-use plastic gloves.
   b. Use a fork to stabilize the product.
   c. Wear nylon or metal mesh gloves.
   d. Hold the meat firmly with a bare hand.
2. When an employee brings prescribed medication to work that must be refrigerated, it should be:
   a. Put in a closed and labeled container in the employee lounge refrigerator.
   b. Placed in the back corner of a refrigerated display case.
   c. Wrapped in plastic and put in the ice machine.
   d. Sent home with the employee because it is not allowed in the area.
3. The Occupational Safety and Health Administration requires all retail food establishments maintain a file with information on how to treat chemical exposures. These documents are:
   a. Material Sample Detail Sheets.
   b. Medical Symptom Data Signs.
   c. Material Safety Data Sheets.
   d. Medical Sample Design Contents.
4. The document that lists when an accident occurs and what action was taken is called an:
   a. Incident report.
   b. Evaluation record.
   c. Accident record.
   d. Insurance form.
5. Every retail food establishment must have the appropriate:
   a. Number of shelves.
   b. Fire extinguishers.
   c. Inventory control system.
   d. Number of checkout stations.

Answers to the multiple-choice questions are provided in Appendix A.

Reference/Suggested Readings

Code of Federal Regulations. 2001. *21 CFR 129 - Processing and Bottling of Bottled Drinking Water* U.S. Government Printing Office. Washington, D.C.

Food and Drug Administration (2001). *2001 Food Code*. United States Public Health Service. Washington, DC.

Food Marketing Institute (1996). *Pre-Emergency Planning and Disaster Recovery Manual*. Loss Prevention Services Department. Washington, DC.

McSwane, D., Rue, N., Linton, R. (2003). *Essentials of Food Safety and Sanitation, 3rd. Edition*. Prentice Hall, Upper Saddle River, NJ.

Learn How To:

- Identify the need for education and training in retail food establishments.
- Locate resources available for food safety and sanitation training including media, government materials, and the Internet.
- Apply evaluation tools to measure training outcomes.

CHAPTER 9

Education and Training

Improper Training Creates a Hazard

At the end of her shift, a three-year employee began to clean out the seafood display case. She removed all the seafood products, the ice was removed and the case was cleaned according to a prescribed procedure. As she left for the day, she told her replacement the case was clean and it just needed to be filled with ice and restocked. Her replacement had just gotten out of training, so she told him to refill the case with ice and replace the product according to a written diagram.

The new employee began to fill the case with fresh ice but there was not enough ice in the machine to fill the case. He saw the buckets of ice she had removed from the case earlier sitting on the floor. Unknowingly, he refilled the case with some of the used ice from the buckets. He replaced the product into the case following the written diagram. He placed the cooked shrimp next to the raw shrimp with no barrier between the two. The diagram merely said, "shrimp."

What hazards now exist in the seafood case? How could this have been avoided?

An effective training program is essential for preparing new employees.

Learners Today Influence Training

As a trainer, you must recognize the age differences of your audience. The work force in today's retail food establishments is quite diverse. Some employees are young and some are older. Some speak English as a first language and some do not. Your training plan should take into consideration the cultural and age diversity of your audience to assure a successful training program.

Today's Mixed Generational Work force				
Born:	**Veterans 1922-1942, 52 million**	**Baby Boomers 1943-1960, 73.2 million**	**GenXers 1960-1980, 70.1 million**	**Nexters 1980-2000, 6.7 million**
Work ethic	Dedicated	Driven	Balanced	Determined
View of authority	Respectful	Love/Hate	Unimpressed	Polite
Training type	• Traditional Classroom.	• Lifelong learning • Scannable material • Read selectively.	• Computer-based • CD ROM • Interactive • Read less.	• Read more • Learn Interactive • Use more technology.

Source: ***Generations at Work*, AMACON 2000**

Current Information and Technical Challenges

Corporate trainers or consultants who contract for work face the challenge of keeping their knowledge base current. Each site also has individual problems related to geographic location, corporate culture, employee turnover, and customer mix. As more businesses consolidate, the smaller "Mom and Pop" operations are becoming less common. Very large retailers are now starting chains of superstores that sell everything, including retail food items. These market trends affect the methods and delivery systems of traditional training.

Training needs are growing. Technology, the Internet, and the latest teaching techniques are working to keep up. Before selecting training materials and techniques you must do your homework. You must consider the needs of your learners as well as the resources available in your operation. Think critically about each training choice you make:

- Research the needs of your operation
- Review prepared training programs

- Compare operational needs to the training materials.

Training is Important

Training is expensive but consider the alternative. Repairing a problem costs time, money, and may cause serious harm to your reputation and sales. The positive result of a well-trained staff cannot be ignored.

Without training time and resourses can be wasted.

Training Benefits and Barriers

Benefits of Training	Barriers to Training
• Improved customer satisfaction	• Perceived lack of time
• Lower employee turnover	• Lack of commitment
• Lower costs	• Lack of knowledge or ability to train
• Fewer accidents	• The culture of the organization
• Better quality	• Lack of a well-thought-out training plan.
• Safer, more efficient employees.	

Create a Plan for Training

Answer the following questions before you develop a training program:

- Which department personnel will attend and what area do they manage, supervise, or work in? This may include:
  - ▲ Meat department
  - ▲ Delicatessen department
  - ▲ Produce department
  - ▲ Bakery department
  - ▲ General store merchandising
  - ▲ Receiving and storage.

- How many learners will be in each group?
- What amount of time is allotted for the program?
- Where will the training take place?
- What teaching and learning resources are available?
- What is the average level of knowledge, skills, and abilities?
- Are there any language differences to be addressed?
- What outcomes does upper-level management expect?
- Are there any current issues that need immediate attention?
- Do evaluation tools exist for this site or will you need to provide them?

Once you have answered these questions, a plan can be developed to conduct training. This approach is aimed at group presentations, usually scheduled for a half-, whole-, or two-day sessions. The length of time spent on training depends on the topic and outcomes expected.

Departments have different needs and materials that require specific instruction on operation and maintenance. Safe food practices—keep it hot, keep it cold, keep it clean—cross all lines. Application varies according to the items within the area. It takes time and patience to teach others how to teach. The motto "each one teach one" is a mainstay in vocational learning.

What do Your Trainees Need to Know?

Communicating with your employees can help you to better understand what training you need. Adults respond better to training that is relevant to their everyday work life. However, there are some things every retail food establishment should train employees about, including:

- Emergency situations and response systems
- Issues related to health and hygiene
- Common causes of foodborne illness
- Food safety management procedures that include time and temperature controls
- Cleaning and sanitation procedures
- How to safely use and clean equipment and utensils
- Accident prevention measures
- Customer relations and workplace etiquette

Orient employees to all safety equipment.

- Company policies and procedures
- Department specific procedures and protocols
- Reportable illnesses and awareness.

This list contains general suggestions. Each trainer must adapt their needs to available resources, time, type of learner, and any other requirements that need to be addressed.

Recommended Times for Training

Short sessions—30 to 45 minutes at a time are easier to handle. Most adults prefer to learn a task or fact and then immediately try it out.

When longer classes are needed, break up the time with activities and interaction that allows participants to get up and move around.

Standard Operating Procedures Federal, State, and Local Rules and Regulations

Every retail food establishment must comply with the rules and regulations of federal, state, and local agencies. Retail food establishments must employ standard operating procedures (SOPs) and good retail practices (GRPs). These procedures and practices must be taught to and regularly practiced by employees. Your training plan will need to include each element employees are responsible for using in their daily work.

Hazard Analysis Critical Control Point System

In the next chapter, you will learn about food safety management programs. If HACCP is a part of the food safety management program, employees will need special instruction in how to use the HACCP food safety system in your establishment. Additional instruction may be needed to assure proper monitoring of critical limits and recording of information to verify the system is working properly.

Set the Stage for Training

Training may take place in a variety of settings and the number of variables within each of those settings can be surprising. Follow these guidelines to assure the comfort of your learner:

- Time of season, temperature of room, and scheduled breaks are all important
- Active learning allows employees to better retain learning
- Discussion questions, demonstrations, sample test questions and illustrations improve participation
- Narrow topics will help the learners stay focused.

Conduct the Training

When the training plan is complete, the next step is to conduct the training. All training should be centered on company policies and procedures. Be sure to consult your procedure manuals prior to conducting any training.

You should always schedule training far enough in advance to be sure everyone has time to plan for it. Try not to cancel any planned training because it will make the training seem less important. An effective training session includes the following points:

- Keep it simple
  - Trying to teach too much at a time can be counterproductive
- Stay focused
  - Set clear objectives for the training session
- Make it fun
  - Use color
  - Use activities
  - Involve the audience
  - Use audiovisual aids as an enhancement, not a replacement for instruction.

Measuring Performance Against the Standards

To make sure learning has taken place, consider the following actions:

- Provide follow-up discussions and reminders when the employee is back on the job. This helps to reinforce the learning that took place in the training session.

- Observe whether or not people perform the job as taught. Watch your employees perform their jobs after the training is completed. Look for the new skills they have learned. When they use their new skill, praise them. When they do not, coach them on the right way.

Good training

Poor training

- If an employee has trouble with a new skill, retrain until he or she does it right.
- Use written tests with caution. Employees who are not literate need special consideration. Verbal questions work better than written tests. Once again, recognize outstanding performance. Also follow up verbal recognition with a hard copy in the employee's record.
- Communicate positive and consistent feedback.
- Provide verbal and written feedback to the employee.
- Ask the employee for suggestions that would help in learning the task.
- Treat everyone with respect. Ridicule and sarcasm do not belong in training sessions.

Training Moments

Learning can occur continuously in retail food establishments.

- Be a positive role model for your employees.
- Use normal happenings as a training moment to improve performance.
- When employees are not practicing proper procedures, be a coach. Teach them the right way to do it.
- Reinforce good habits. Reward employees who perform their jobs correctly.

Check it Out

*Back to the Story... **The employee who restocked the seafood case created several problems. By placing raw and cooked shrimp side by side without a barrier, the cooked product will have to be discarded. Cross contamination from the raw to the cooked shrimp introduces the possibility of disease-causing microbes. Using half-melted ice that was in the case with all types of seafood cancels out the benefits of cleaning the case. An employee who does not understand basic cleaning and sanitizing principles can cause severe damage to the food safety program.***

- ✓ As a trainer, be clear on your standards for performance.
- ✓ Target your training, plan for training, and take time to keep up with our changing, growing industry.
- ✓ The principles of good training include:
  - Keep it simple
  - Plan for activities, content, and breaks
  - Make it fun
  - Involve the learner
  - Use visual aids.

Quiz 9 (Multiple Choice)

Please choose the BEST answer.

1. Today's food industry leaders understand training is:
   a. A cost without benefits.
   b. Not necessary.
   c. An investment.
   d. A one-time event.
2. One of the benefits of effective, consistent training is:
   a. Higher turnover.
   b. Fewer accidents.
   c. Better quality.
   d. b and c.
3. What, who, how, when, and where are questions that need to be answered in which of the following?
   a. Deciding on what training materials are needed.
   b. Preparing to train employees.
   c. Evaluating training.
   d. All of the above.
4. A goal(s) in a great training session is:
   a. Make it difficult.
   b. Make it fun.
   c. Do not use visual aids.
   d. Make it passive.
5. To assure the transfer of learning, one thing to consider is:
   a. Providing follow-up discussions.
   b. Catching people doing it wrong.
   c. Serving refreshments in class.
   d. Providing training at the end of the shift.

Answers to the multiple-choice questions are provided in Appendix A.

References/Suggested Readings

Zemke, R., Raines, C., Filipzpak, R. (2000), *Generations At Work.* American Management Association. New York, NY.

Dolasinski, Mary Jo (2004), Training the Trainer: Performance-based Training for Today's Workplace. Prentice Hall, Upper Saddle River, NJ.

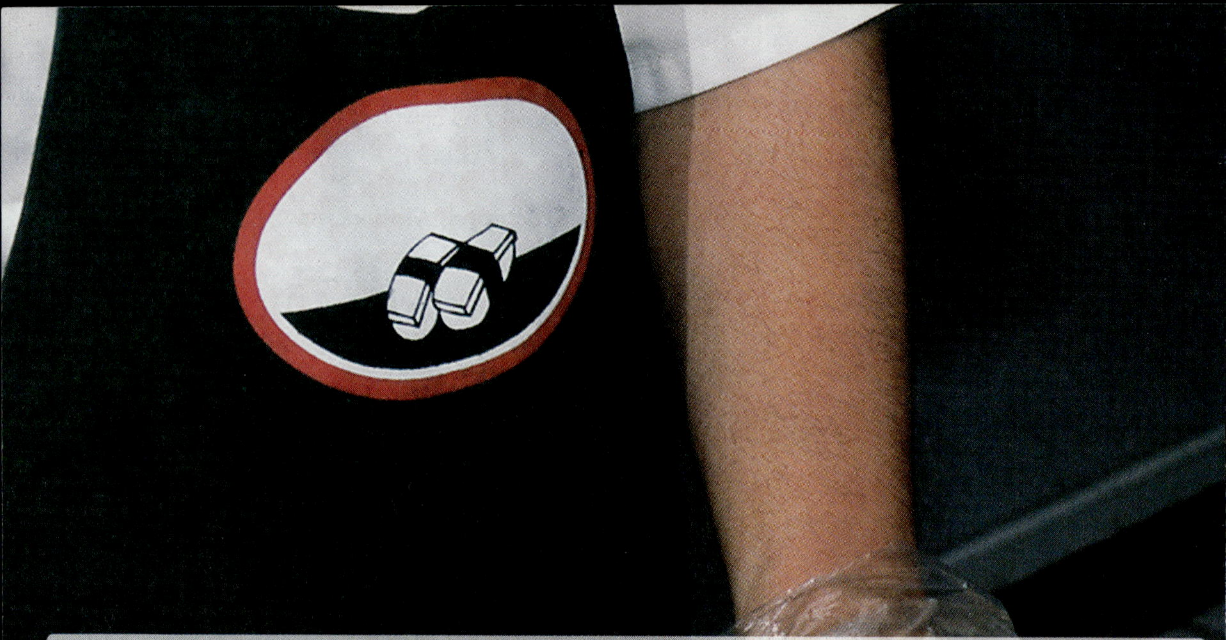

Learn How To:

- Recognize how food safety management programs in retail food establishments reduce the risk of foodborne illness.
- Apply the Hazard Analysis Critical Control Point (HACCP) system for food safety management.
- Define Critical Control Point (CCPs) and identify critical limits as used in the HACCP system.
- Determine the roles of managers and employees related to food safety management.
- Comply with a food recall procedure.
- Identify the need and function of product identification and product coding.

CHAPTER 10

Food Safety Management Programs

Potato Salad Encounters a Critical Control Point Violation

A large order was received from the central commissary including foods that needed to be stored under refrigeration, frozen, and in dry storage. During the storing process, some packages of potato salad were accidentally stored on a shelf in the storeroom. The potato salad needed to be kept under refrigeration.

Sometime later, an employee who was organizing the dry storage area discovered the packages that were stored out of refrigeration. The employee took the potato salad to the walk-in refrigerator and placed it under cold storage. The employee never took the temperature of the potato salad. Later, another employee pulled the potato salad from the walk-in cooler and placed it in the refrigerated display case in the deli area where it was sold to customers.

What are the potential problems from this incident? What important critical control point was ignored?

The Problem

The challenge of reducing the number of cases of foodborne illness has made it necessary to reevaluate our nation's food safety system. The retail food industry and food regulatory agencies must deal with:

- New hazards that cause foodborne illness
- A global food supply that carries more risks
- New techniques and equipment for processing, displaying, and serving food
- A change in eating patterns by customers including more ready-to-eat products
- An increase in the number of people who are highly susceptible for foodborne illness due to age and condition.

The size and scope of the food industry increases the workload on the Food and Drug Administration (FDA), U.S. Department of Agriculture (USDA), and state and local food regulatory agencies. The FDA identified more than 30,000 food manufacturers and 20,000 food warehouses that process and store products. Inspection of all these operations is nearly impossible. The retail and foodservice portion of the food industry is even larger and consists of more than one million establishments with a work force of over 15 million people.

Food safety and food quality are both major concerns for retail food establishment managers and their staff. Food safety focuses on assuring the products sold are safe to eat. Regulatory standards and guidelines related to safe food handling at retail levels are discussed in this book and in the *FDA Food Code*. Food quality is also very important. High quality often equals high profits and returning customers. With an average profit margin of 1-2%, it makes good business sense to produce and sell the safest and highest quality food possible.

The goal of food safety management is to prevent foodborne illness.

The Solution

Chapter 3 identifies the major risk factors related to foodborne illness. The most important factors are:

- Improper holding temperatures
- Poor personal hygiene
- Contaminated equipment
- Inadequate cooking
- Food from an unsafe source.

Food Safety Management Programs Involve:

- Controlling time & temperature
- Practicing good hygiene
- Preventing cross contamination
- Cleaning and sanitizing properly.

A food safety management program should protect foods from the factors known to lead to foodborne illness. To assure food safety, we need to focus on the following four keys to foodborne illness prevention:

- Time and temperature control
- Good personal hygiene practices and procedures
- Elimination of cross contamination
- An effective cleaning and sanitation program.

Food safety management programs in retail food establishments must be customized to fit company policies, the size of the establishment, and the number and types of customers served. To operate a food safety management program, you must understand the hazards that cause foodborne illness and how these hazards can be prevented. Focus the program on elimination or control of the risk factors known to contribute to foodborne illness.

The food safety procedures in your management system must be easy for your employees to understand and implement. Management in your operation will be up to you.

Start with a Good Set of Instructions

The ability to do a job correctly depends on good instructions and properly trained employees. This holds true for all functions in a retail food establishment including marketing, purchasing, operations and customer services. This includes how to:

- Break down pieces of equipment for cleaning
- Stock and rotate products on shelves
- Wash hands properly
- Operate the slicer in the deli or meat department
- Take care of an emergency.

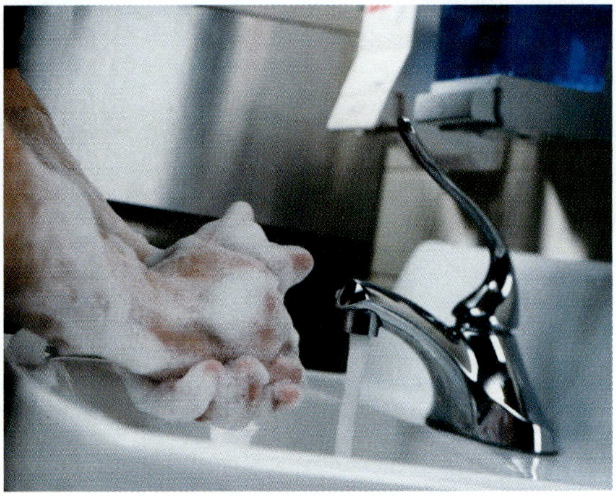

Put safety procedures in place to prevent foodborne illness.

In retail food establishments, there are many different types of procedures that need to be done correctly to protect the safety of foods. A detailed step-by-step form of instruction and training will help the employee do a job properly.

There are many methods the retail food industry uses for this type of information. In some cases, a book of procedures is used. In other cases, procedures are posted near workstations for easy access by employees. There are also many terms that can be used for these procedures. **"Best Practices"** and **"Standard Operating Procedures" (SOPs)** are two terms commonly used to describe written procedures on how to perform a job. It does not matter what these procedures are called so long as they are written clearly, explained to employees, and performed correctly.

There are many different types of SOPs that can affect the safety and quality of foods during receiving, storage, preparation, and sale.

Some examples of SOPs used by retail food establishments include:

- Good personal hygiene programs
- Cross contamination control
- Cleaning and sanitizing programs
- Pest management
- Storage practices of foods and ingredients
- Water and ice handling
- Use of chemicals.

The Food Marketing Institute (FMI) has provided model SOPs for molluscan shellfish handling in a document titled "*Molluscan Shellfish Handling for Retailers (FMI, 2001b).*" In this document, six important elements are indicated for each SOP including:

- Title of the task to be completed
- Date the SOP was issued
- Area or department to use the SOP
- Detailed, specific actions that need to be taken to complete the task
- Any documentation that needs to be completed to verify the task was accomplished
- Approval signatures.

"Best Practices" and "Standard Operating Procedures" (SOPs) are written procedures on how to perform a job.

Following is an example of an SOP from this document specific for personal health and hygiene practices. Note the SOP contains a very specific list of instructions and includes directions for monitoring and corrective action. In this SOP, "critical items" are identified as those steps in the procedure most critical to food safety.

SOP for Personal Health and Hygiene

Background Notes Date Effective 4/5/01

- This procedure outlines personal health, hygiene and handwashing requirements required for the safe handling and sale of foods
- Employees diagnosed with a food-related illness or showing symptoms of an illness are not permitted to work with food, food-contact equipment, utensils or surfaces
- Employees with festering cuts, abrasions, boils, or any condition that may cause flaking or exposed skin may not have contact with food or food-contact items.

Departments: All Departments

Specific Instructions

- Critical Item - Only healthy employees are permitted to handle food, food-contact equipment, and utensils. Employees that have been diagnosed with a food-related illness or are experiencing symptoms (sore throat with fever, diarrhea, vomiting or jaundice) must immediately notify their supervisor and remove themselves from the food environment.
- Critical Item - Employees with festering cuts, abrasions, boils or any other condition that may cause flaking of exposed skin must immediately notify their supervisors and remove themselves from the food environment. An infected wound must be freshly and properly bandaged and covered with an impermeable cover (surgical glove), which will prevent contact with food or food-contact items.
- Bathe daily and have clean, well-groomed hair. Present a clean appearance.
- Wear clean, washable outer garments, clean apron and clean shoes. Safety shoes may be necessary.
- Wear clean, effective hair restraint.
- Have clean hands and fingernails. False fingernails and nail polish are not permitted.
- Employees must avoid wearing any jewelry. Plain wedding bands and medical identification bracelets may be exempt.
- Critical Item - All employees must wash their hands and exposed portions of their arms thoroughly:
  - Before starting work.
  - After using the restroom.
  - After eating, drinking, or using tobacco products.
  - After touching their mouth or anything that has been in their mouth.

Model SOP for Personal Health and Hygiene for Molluscan Shellfish Handling (Used with permission from the Food Marketing Institute - FMI, 2001b). Contined

SOP for Personal Health and Hygiene (continued)

e. After touching their hair, nose, ears, etc.

f. After handling trash, garbage, dirty utensils or soiled equipment.

g. After sneezing or coughing into their hands, or using a handkerchief or disposable tissue.

h. After changing from raw to cooked or ready-to-eat foods.

i. After any absence from the work area.

- Critical Item - Washing hands and exposed portions of arms is accomplished by:
  a. Vigorously rubbing hands together, lathering soap, using hot water for at least 20 seconds.
  b. Rinsing with clean water.
  c. Drying with disposable paper hand towels, clean continuous dispenser towels or air dryers.
- Avoid bare hand contact with any ready-to-eat foods, including shellfish. Use single-use gloves, deli tissues or utensils to handle product. Change single-use gloves frequently to prevent contamination. Remember gloves are not a substitute for proper hand washing.
- Remove personal items from pockets where they may fall into food, such as cigarettes, lighters and pens.
- Employees may only eat, drink, smoke or chew (tobacco or gum) in designated break areas.
- Employee street clothing, personal belongings or personally owned foods and beverages should be stored in designated areas.

Monitoring

Daily: Department manager should visually observe employees for:

- Symptoms of illnesses or skin infections
- Proper handwashing practices and frequency.

Corrective Action

- Refer all employees diagnosed with a food-related illness, displaying symptoms of illnesses or skin infection immediately to department manager
- Refer employees to repeat handwashing procedures.

Model SOP for Personal Health and Hygiene for Molluscan Shellfish Handling (Used with permission from the Food Marketing Institute - FMI, 2001b).

The HACCP Approach to Food Safety

In the food industry, there are a variety of programs that can be used to manage food safety. **Hazard Analysis Critical Control Point (HACCP)** is a prevention-based program used by many segments of the food industry to assure food safety. The HACCP system can control hazards such as disease-causing microorganisms, chemical substances and physical objects that are common causes of foodborne illness (refer back to Chapter 2 for a detailed description of these hazards and their control).

In 1959, the National Aeronautic and Space Agency (NASA) approached the Pillsbury Company to produce food safe for astronauts to eat in zero gravity conditions in the space capsules. The project goal was to develop a system that assured nearly 100% freedom from contamination by microbial, chemical, or physical hazards. This effort resulted in the beginning of the HACCP concept for food safety management.

In food manufacturing industries, HACCP has been required for many years. Currently, HACCP is not required for retail food establishments except in a limited number of states and jurisdictions. However, many retail food establishments have implemented HACCP programs voluntarily to assure the safety and wholesomeness of the products they produce and sell.

When a HACCP program is developed, a team is chosen to discuss two fundamental questions:

1. Are there any hazards or risks associated with the ingredients, the process, or the finished food?
2. How can I make sure the identified hazards and risks are kept at a safe level?

The Seven Principles in a HACCP System

The basic structure of a HACCP system consists of seven principles shown in the following chart. A HACCP team identifies the potentially hazardous ingredients and foods used and sold at the establishment. The HACCP team also discusses the different processes involved to prepare these items. After implementation of the seven principles, individuals must be trained and the program must be evaluated to assure employees understand their roles and responsibilities related to HACCP.

Seven Principles in a HACCP System

1. Hazard analysis.
2. Identify the Critical Control Points (CCPs) in food preparation.
3. Establish Critical Limits (thresholds) that must be met at each identified CCP.
4. Establish procedures to monitor CCPs.
5. Establish the corrective action to be taken when monitoring indicates a Critical Limit has been exceeded.
6. Establish procedures to verify the HACCP system is working.
7. Establish effective record keeping that will document the HACCP system.

Principle 1 - Hazard Analysis

In this first step, the team identifies hazards associated with ingredients that are received, those that may be associated with each step in the flow of food, and those important at the time of purchase or consumption by the customer. Hazards are then assessed for their risk. Risk indicates a level of danger based on how likely the hazard is to occur and how severe the hazard could be from a public health standpoint. Those hazards most likely to occur or pose the highest risk of disease will be addressed in the HACCP system.

Principle 2 - Identify Critical Control Points or CCPs

The *FDA Food Code* defines a CCP as a point at which loss of control may result in an unacceptable health risk. A CCP can be thought of as a point in the flow of food where a preventative measure can be put into place to control the identified hazards. The team identifies areas or points in the flow of a food product that must be controlled. These are the CCPs.

Principle 3 - Establish a Critical Limit for Each CCP

Critical limits are the upper and lower boundaries of food safety and are set at each CCP to reduce hazards and risks to safe levels and assure the CCP is under control. Critical limits are usually based on time, temperature, pH, and moisture content of a food. When you set a critical limit, the goal is to prevent or eliminate the hazards identified during the hazard analysis portion of the program. A critical limit should be as specific as possible, such as "ground beef must be heated to an internal temperature of 155°F (68°C) or above for at least 15 seconds." A well-defined critical limit makes it clear and easy to determine when the limit has not been met. In retail HACCP plans, critical limits for time and temperature are most often used to keep hazards under control.

Principle 4 - Monitoring CCPs

Critical Control Points and critical limits are only effective if they are monitored during the flow of food. Monitoring includes activities such as measuring temperature and time. Monitoring helps to assure the processes involved in the flow of food are done correctly and the hazards are in control.

Principle 5 - Take Corrective Action

Whenever the critical limit for a CCP is exceeded, something must be done to assure the food is safe. When a critical limit is not met, fix it. Corrective action may include adjusting product temperature or time, or even discarding the product if necessary to assure the safety of your customers.

Principle 6 - Establish Procedures to verify the HACCP System Is Working

Once the HACCP system is in place, a designated process is used to make sure it is effective. The *FDA Food Code* recommends three phases to verification including 1) scientific verification of CCPs and CCP limits, 2) assuring the HACCP plan is functioning effectively, and 3) repeated validation of the HACCP plan by internal or external food safety audits.

Principle 7 - Establish Effective Record Keeping.

This principle enforces the importance of documentation of CCPs, CCP limits, and monitoring activities. These records are the only proof the process is in control and the retail food establishment is complying with their HACCP plan.

Retail HACCP: A Process Approach

As the food industry develops and implements HACCP programs, it makes choices between a HACCP program that is "product"-specific or one that is "process"-specific.

- A **product-specific** HACCP program is tailored to the production of a specific food product. When product-specific HACCP programs are used, a HACCP plan is developed for each food product produced. This approach works well, especially in food manufacturing plants, where a limited number of products are handled.
- A **process-specific** HACCP program is customized to the processes used to produce a particular food or group of similar foods. In the retail food industry, numerous types of potentially hazardous foods are produced and sold within a single retail food establishment. It often does not make sense and is not manageable to develop a HACCP plan for each one of these food products.

The retail food industry frequently relies on process-specific HACCP programs to assure food safety. Retail food establishments base their HACCP plans on the process, or flow of food, rather than the actual product produced. This is called a process approach to HACCP. It simplifies the procedure so HACCP plan development is manageable. For example, for hot-holding, it does not matter what type of potentially hazardous food we are monitoring. In all situations, the key is to keep the food temperature above 140°F (60°C). The product is taken into consideration only when there is a difference in temperature guidelines such as cooking [i.e., ground beef must be cooked to at least 155°F (68°C) while poultry must be cooked to at least 165°F (74°C)].

Every process in the flow of food is created from a few basic steps. They may include:

- Received and stored frozen
- Thawed
- Received cold
- Stored cold
- Prepared cold
- Displayed cold
- Cooked
- Hot-held
- Displayed hot
- Cooled
- Reheated
- Service and handling.

If we take the process approach, we know there are certain temperature recommendations for each of these steps. We will follow these temperature guidelines in our food safety management system.

Some of the steps in the flow of food:

Receiving

Service and handling

The figure below shows the different areas in the flow of food and the temperature guidelines used. These guidelines provide the basis for developing safety limits for the HACCP system.

Regardless of the type of food handled and sold in a retail food establishment, the food flow is made up of a series of steps. For most of the steps in the flow of food, time and temperature control is the same regardless of the potentially hazardous food. There are a few exceptions. In receiving, eggs and reduced oxygen packaging (ROP) foods may be received at a temperature other than 41°F (5°C). ROP foods may also have to be held cold at a temperature other than 41°F (5°C). Safe cooking temperatures will also differ depending on the product being cooked.

FLOW OF FOOD STEP	TEMPERATURE GUIDELINES FOR POTENTIALLY HAZARDOUS FOODS (PHFS)
Frozen Foods	
Receiving	• Foods are solidly frozen when received
Storage	• Keep all foods solidly frozen
Thawing	• Thaw in the refrigerator at 41°F (5°C), as part of the cooking process, under cool running water, or in a microwave oven.
Cold Foods	
Receiving	• Receive all potentially hazardous foods (PHF) at 41°F (5°C) or below • Whole shell eggs at 45°F (7°C) ambient temperature or below • Dairy products at 45°F (7°C) or below • Shucked shellfish at 45°F (7°C) or below • Live shellstock at 50°F (10°C) or below • Reduced oxygen packaged (ROP) foods may have special temperature requirements.
Refrigerated Storage Cold-Holding/ Display	• Store all PHFs at 41°F (5°C) or below • ROP foods may have special temperature requirements

(Continued)

FLOW OF FOOD STEP	TEMPERATURE GUIDELINES FOR POTENTIALLY HAZARDOUS FOODS (PHFS)
Preparing Cold Foods	**(Continued)**
	• Minimize time between 41°F (5°C) and 135°F (57°C) to less than 4 hours for PHF.
Preparing Hot Food	
Cooking	Cook to the safe product temperature: • Poultry and stuffed meats - 165°F (74°C) or above for 15 seconds • Ground beef and pork and game animals - 155°F (68°C) or above for 15 seconds • Shell eggs, beef and pork (other than roasts) and fish - 145°F (63°C) or above for 15 seconds • Beef roast (rare) 130°F (54°C) for 112 minutes or 140°F (60°C) for 12 minutes • Beef roast (medium), pork roast, and ham - 145°F (63°C) or above for 3 minutes • All other PHFs - 145°F (63°C) or above • Microwave cooking - 165°F (74°C) or above, allow to stand for 2 minutes after cooking.
Hot-holding/ Display	• Hold all PHFs at 135°F (57°C) or above.
Cooling	• Cool all foods from 135°F (57°C) to 70°F (21°C) within 2 hours and from 135°F (57°C) to 41°F (5°C) within 6 hours [total cooling time should not exceed 6 hours from 135°F (57°C) to 41°F (5°C)].
Reheat	• Reheat all PHFs to greater than 165°F (74°C) within 2 hours.
Preparation/ Handling/ Service	• Restrict the amount of time potentially hazardous foods are held between 41°F (5°C) and 135°F (57°C) to 4 hours or less. **(Source: *FDA Food Code*)**

HACCP plans are often shown as a flow diagram. A flow diagram provides a visual map of the different ingredients and each of the steps involved in the handling and preparation of food. Flow diagrams may also include the CCPs included in the HACCP plan. The figure below provides an example of a flow diagram for a cooked and hot-held product (rotisserie chicken). For this product, cooking and hot-holding are identified as critical control points (CCP). The critical limits for this product are cooking the chicken to an internal temperature of at least 165°F (74°C) for 15 seconds and holding it at 135°F (57°C) or above. Cooking is important for destroying any harmful bacteria that may have been present on the raw chicken. Control of hot-holding prevents growth of harmful bacteria that may have survived cooking or that contaminated the chicken after cooking.

A. Rotisserie Chicken

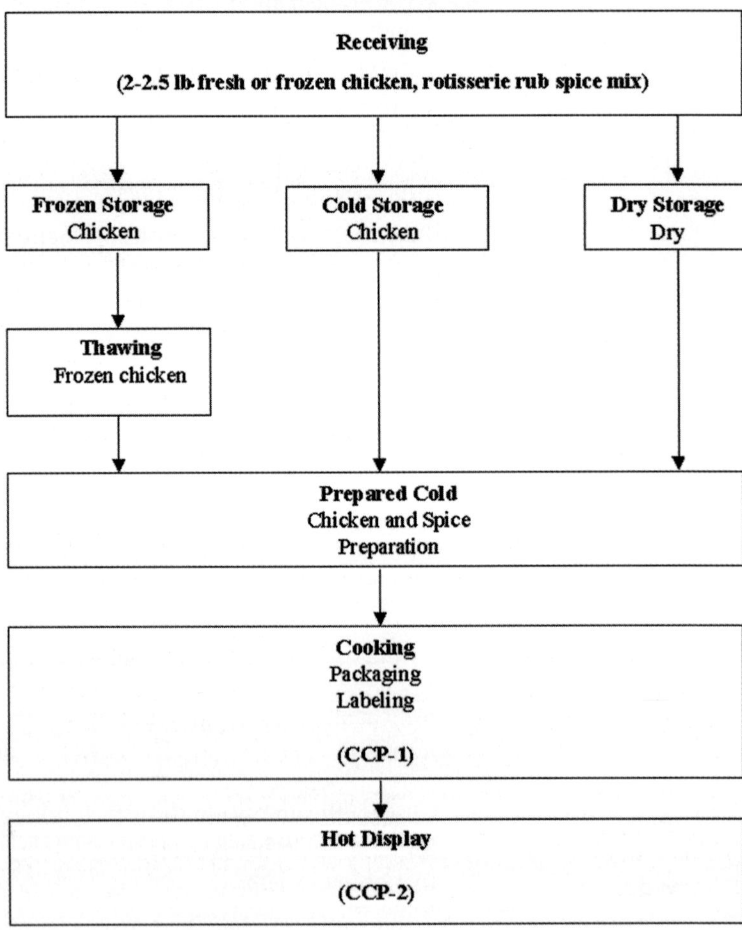

Rotisserie chicken flow diagram with CCPs identified

Food Safety Management and HACCP Requirements

A growing number of state and local jurisdictions require retail food establishments have a HACCP system in place. These agencies also conduct inspections of retail food establishments using HACCP principles. You may contact your local jurisdiction to obtain information about developing a HACCP system for your establishment.

The federal government requires HACCP plans and inspections for food processors that manufacture certain varieties of meat, poultry, and seafood products and unpasteurized juice products. A HACCP program is recommended when reduced oxygen packaged (ROP) foods are produced in retail food establishments.

Other Food Safety Management Programs

HACCP can be a very effective program but it is not the cure-all for all food safety problems and in all segments of the food industry. There are many other types of approaches that can be used.

In the food processing industry, a program called Good Manufacturing Practices (GMPs) is used as part of the food safety management program. GMPs refer to minimum sanitary and processing requirements necessary to assure the production of safe and wholesome food. GMPs include provisions for food handling related to personnel, building and facilities, equipment and utensils, and production and process controls. When implemented properly, GMPs help control the possibility of contamination from poor personal hygiene, chemicals, and physical objects. They also help assure cleaning and sanitizing are done properly. There are FDA requirements for GMPs in food manufacturing operations, and they are listed in Section 21 of the *Code of Federal Regulations*, Part 110.

More recently, on-farm production of foods has adopted a new term called Good Agricultural Practices (GAPs). The goal of GAPs is to reduce the risk of contamination on farms mainly for fruit and vegetable products. Programs that use GAPs focus on farm management programs such as personal hygiene, irrigation water, manure, and cleaning procedures for freshly harvested produce. Assuring your produce vendors adopt GAP type programs will be very important for reducing the risk of ready-to-eat produce displayed and sold in your establishment.

Following along the same pathway, the Food Marketing Institute has embraced a program called Good Retail Practices (GRPs). GRPs are similar to the GMPs and

GAPs used in food manufacturing and food production, respectively. The main difference is GRPs focus on food safety practices for retail food distribution, preparation, display and sale. In many retail food establishments, GRPs are considered an important and very effective way to manage food safety. GRPs can be flexible and written to enhance product quality and shelf life. Some examples of GRPs include:

- Receiving guidelines
- Storage guidelines
- Preparation guidelines
- Packaging and labeling guidelines
- Display guidelines
- Personal health and hygiene
- Employee and customer education
- Record keeping.

The Food Marketing Institute has published a document on sushi preparation called *FMI's recommendations for the safe preparation of sushi (FMI 2001a)*. In this document, several examples of GRPs are provided. Examples of receiving and storage GRPs, respectively, from this document are provided in this example.

Examples of GRPs

Product labeling

Display guidelines

GRP Example

A. Receiving Raw Sushi-grade Fish

All ingredients should come from approved sources. Seafood products identified as a parasite hazard in the Seafood Hazard Guide should be obtained from a supplier who has frozen the product throughout to -4°F (-20°C) or below for seven days, or -31°F (-35°C) or below for 15 hours. If the species is identified as a histamine producer, documentation from the supplier should show temperature control from harvest through delivery was sufficient to prevent any stage of decomposition.

General Receiving Instructions:

- Check the incoming product for date code, physical condition of packaging, physical defects, insect infestation, and food quality
- Check the temperature to assure the product is frozen solid or has been refrigerated to 41°F (5°C) or below
- Place ready-to-eat foods in the cooler immediately
- Place frozen product delivery into the freezer immediately
- Rotate delivery of the product to facilitate the first-in, first-out method of stock rotation
- Obtain a letter of guarantee from the supplier stating all sushi-grade fish has been frozen at -4°F (-20°C) or below for at least seven days, or -31°F (-35°C) or below for 15 hours before it is released for distribution
- Obtain a letter from the supplier verifying that, based on microbiological testing, the fish is high sushi-grade
- Reject products that do not meet the above criteria.

(Continued)

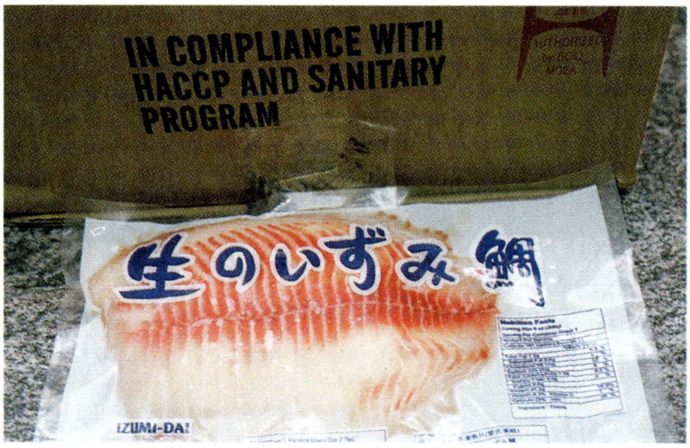

GRP Example **(Continued)**

B. Storage Guidelines for Raw Sushi-grade Fish

To maintain product quality once the retailer has accepted shipment:

- Place product in the appropriate storage unit.
- Equip refrigeration unit with a calibrated thermometer easily accessible for viewing.
- Use recording devices and alarm systems.
- Avoid cross contamination.
- Store product in a specific, predesignated area in the refrigeration unit, away from other raw and cooked foods.
- Keep product containers 6 inches above the floor, away from walls and ceilings, and preferably on non-wooden supports, shelving, or surfaces.
- Use an inventory system to make sure the first product in is the first product out (FIFO).
- Check the temperature of the refrigerator and freezer units at least twice daily to assure they are operating at 0°F (-18°C) and 41°F (5°C) or below respectively.
- Notify store manager or person in charge immediately if improper temperatures are detected at times other than the unit's defrost times. Remove food from defective unit and place it in one that works properly.
- Store raw foods below or away from ready-to-eat foods.

GRPs for the safe preparation of sushi for storage (Used with permission from the Food Marketing Institute - FMI, 2001a).

Food Recall Procedures

A retail food establishment has limited control over the food they sell. It makes good sense to assure you have checked out your supplier to determine you are receiving the best products possible. Once in the store, your job is to follow good food-handling practices to keep products safe and in good quality. On occasion, however, contaminated products may be received from food manufacturers, slaughterhouses, or farms that can present a hazard to your customers. In these situations, the company that supplied you with contaminated product may put a food recall in place. If this happens, your job is to remove the contaminated product from your shelves so your customers cannot purchase it. Another important part of your food safety management system is to develop a food recall procedure.

A food recall is the system used to protect customers from using a product that may be unsafe. There are different types of recalls that can be issued depending on how much of a risk the product may present. Chapter 11 describes these different levels of food recalls. Most of the time, a recall involves removing the product from retail shelves so it cannot be sold. It may also involve retrieving product already sold or providing information to customers that have already purchased the product.

In the event of a food safety emergency or other emergencies involving security and loss prevention issues, FMI members can use the 24-hour Crisis Notification System to contact FMI staff after regular working hours.

Coding and Product Identification

When monitoring is done, especially for time and temperature, it is important you be able to identify which product was actually measured. You need a procedure to identify the product. This is usually done with a process known as product coding.

Product coding will be different in almost every retail food establishment because it needs to be customized to your in-house operations. One of the more common uses of product coding is to assure foods are displayed and used on a First-in-First-out (FIFO) basis and products with an expired shelf life should be discarded.

Date and source codes are printed on an item to indicate its shelf life. Date codes simplify rotation and help prevent the sale of off-quality products.

Challenges of Food Safety Management

An effective food safety management program ensures the safety of the products you sell and is MANAGEABLE for you and your employees.

There are a tremendous number of challenges related to managing food safety in retail food operations. In the past decade, there has been a debate over how retail food safety programs should be developed and implemented. Like many within the foodservice and food industry, the retail food industry has a high employee turnover rate. There are language and cultural barriers that make food safety education difficult. The retail food industry and regulatory community do not always agree on food safety practices. Even with all these challenges, retail food safety management must be a high priority.

As you develop a food safety management program for your establishment, use the following guidelines:

- **Focus on risk:** Inspection procedures and audits for food safety management programs should focus directly on controlling risk factors that affect food safety.
- **Make it manageable:** The food management program needs to be user friendly. It should be easy to implement and manage. Food safety management programs need to be flexible and related to specific operational procedures. One size does not fit all.
- **Strengthen communication:** The retail food industry and regulatory professionals share a common goal to protect public health by only allowing the sale of safe food. With the same goal, the food industry and regulatory professionals should partner and work together. Education and training need to be encouraged for members of the retail food industry and for regulators.

Check It Out

Back to the Story… ***If you implement a food safety management system, you can avoid situations like the one described in the story at the beginning of this chapter. In the situation with the potato salad, the time and temperature logs were not compared between departments. The potato salad was exposed to unsafe temperatures for too long a time period. Because of the exposure to unsafe time and temperature, the product should***

have been discarded because of the chances of bacterial growth. This type error can be prevented when a good food safety management program is put in place and the employees are properly trained. The food safety management plan would call for monitoring the temperature and setting critical limits for storage of the potato salad. In this case, the food safety management plan would identify the appropriate corrective action (discarding the product) so customers were not put at risk.

- ✓ When a food safety management system is properly followed, it protects against many types of mishandling.
- ✓ Food safety management systems are designed to prevent foodborne illness.
- ✓ Critical limits set the range of safety, for certain conditions like time and temperature, for handling potentially hazardous products.
- ✓ Teach employees how to use the system and then monitor the application to assure food safety.
  - Employees must be encouraged to contact their supervisor if they discover the prescribed critical limit is not being met.
  - Management must then take the steps needed to correct the problem and bring the production process under control.

Quiz 10 (Multiple Choice)

Choose the BEST answer for each question.

1. A risk as used in connection with a HACCP system is:
   a. Failure to meet a required critical limit for a critical control point.
   b. An estimate of the likely occurrence and severity of a hazard.
   c. Greatest for nonpotentially hazardous foods.
   d. Lowest when complex recipes are required to produce a food item.

2. Which statement about HACCP programs is false?
   a. The HACCP system attempts to anticipate problems before they happen and establish procedures to reduce the risk of foodborne illness.
   b. The HACCP system targets the production of potentially hazardous foods from start to finish.
   c. Records generated by the HACCP system can also be used to aid in foodborne disease investigations.
   d. A HACCP system should only be implemented by public health officials who have been certified by the FDA to conduct such programs.
3. Which of the following is an example of a common critical control point in a HACCP plan?
   a. Washing fruits and vegetables.
   b. Cooking hamburgers.
   c. Handwashing.
   d. Inspecting procedures for quality.
4. Which of the following is an example of a critical limit?
   a. Poultry and eggs are purchased from approved sources.
   b. Chicken and noodles are heated on the stove until the center of the product reaches 165°F (74°C) for 15 seconds.
   c. Only pasteurized milk is used by the establishment.
   d. The cutting board is washed and sanitized between chopping carrots and celery for the garden salad.
5. A food recall is:
   a. A procedure for removing unsafe foods from retail shelves.
   b. A process for labeling food products during preparation.
   c. Only important for very large retail food operations.
   d. Only important for USDA inspected foods.

Answers to the multiple-choice questions are provided in Appendix A.

References/Suggested Readings

Bryan, Frank L. 1990. *Hazard Analysis Critical Control Point (HACCP) Systems for retail food and restaurant operations.* Journal of Food Protection 53(11): 978-983.

Bryan, F. L. 1992. *Hazard Analysis Critical Control Point Evaluations: A Guide To Identifying Hazards and Assessing Risks Associated with Food Preparation and Storage.* World Health Organization. Geneva, Switzerland.

Bryan, F.L., C.A. Bartleson, C.O. Cook, P. Fisher, J.J. Guzewich, B.J. Humm, R.C. Swanson, and E.C.D. Todd. 1991. *Procedures to Implement the Hazard Analysis Critical Control Point System.* International Association of Milk, Food and Environmental Sanitarians, Ames, IA, 72 pp.

Corlett, D.A. and M.D. Pierson. *HACCP Principles and Applications*, Van Nostrand Reinhold, New York, 1992

Federal Register, Vol. 59, No. 149. U.S. Government Printing Office, 1994. Washington, DC.

Food and Drug Administration. 2001. *2001 Food Code.* U.S. Public Health Service. Washington, DC.

Food Marketing Institute. 2001a. *FMI's Recommendations for the Safe Preparation of Sushi.* Food Marketing Institute. Washington, DC.

Food Marketing Institute. 2001b. *Molluscan Shellfish Handling for Retailers - A Total Food Safety Management System Guide for Molluscan Shellfish.* Food Marketing Institute. Washington, DC.

LaVella, B. and J. L. Bostic. 1994. *HACCP for Food Service—Recipe Manual and Guide.* LaVella Food Specialists. St. Louis, MO.

J. K. Lotkin. 1995. *The HACCP Food Safety Manual.* Wiley Press. New York, NY.

Pierson, M. D. and D. A. Corlett, Jr. 1992. *HACCP Principles and Applications.* Van Nostrand Rheinhold. New York, NY.

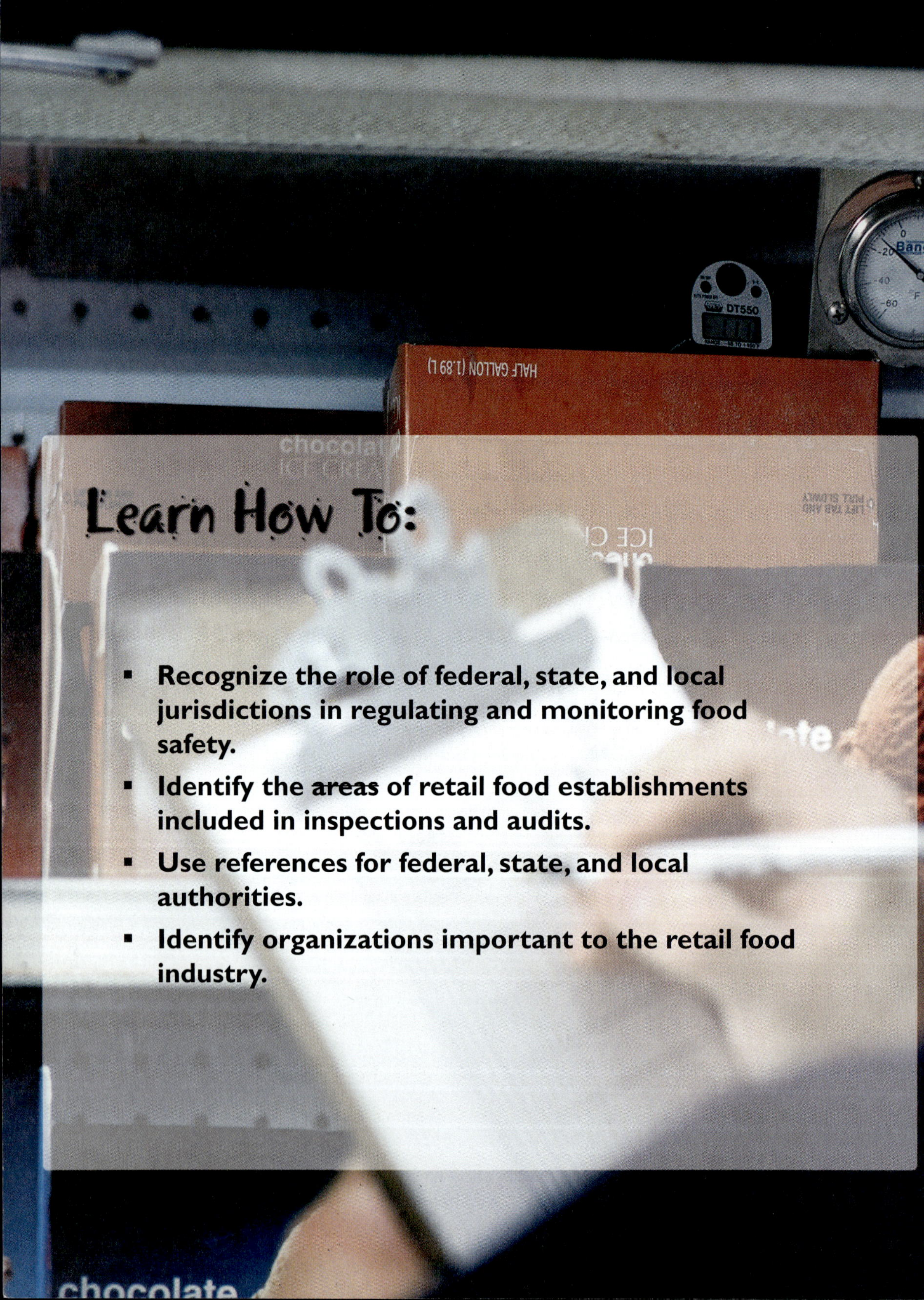

Learn How To:

- Recognize the role of federal, state, and local jurisdictions in regulating and monitoring food safety.
- Identify the areas of retail food establishments included in inspections and audits.
- Use references for federal, state, and local authorities.
- Identify organizations important to the retail food industry.

CHAPTER 11

Food Safety Regulations

Chocolate Candies Recalled

One young boy purchased a box of candies at the local convenience store. He gave them to his mother a few days later as a birthday gift. As she was eating the candy, his mother appeared surprised and fearful to find peanuts in the candy. Immediately after eating the candy, she developed hives and a severe rash and was later rushed to the hospital for care. She was experiencing an allergic reaction due to the peanuts contained in the candy.

Although she knew she was allergic to peanuts, she felt comfortable eating the candy because it did not list peanuts on the ingredient list on the label. The boy's family quickly called the manufacturer to notify them of the problem. The manufacturer quickly ordered a nationwide recall and initiated their recall procedure.

What is your role as a supervisor in response to a food product recall order?

State and Local Regulations

State and local agencies responsible for enforcing food and safety requirements may be under the department of environmental health, public health, or agriculture. Jurisdiction varies from state to state. Contact your local authorities to find out what agency is in charge of food safety in your area.

Most retail food establishments will work closely with their local health department. This agency can provide you with a copy of its current food safety code. Your state and local code contains the food safety standards and regulations that would apply to your type of operation. Familiarize yourself with the key food safety provisions of the code.

Permit to Operate

It is unlawful to operate a retail food establishment unless you have a valid permit issued by the regulatory authority in your area. When a permit is obtained, post it in a prominent location. Permits are generally not transferable from one person to another, from one retail food

Retail Food Market

Monroe County Health Department
Bloomington, IN 47404-3989

Name: ______
Address: ______
City/ State ZIP: ______

Owner: ______

Having complied with the rules and regulations of the Monroe County Health Department as authorized by the Indiana Administrative Code and the Board of Commissioners of Monroe County, Indiana, is hereby authorized to operate a Retail Food Market at the above location for the calendar year ending December 31st:

Issued ______

By ______

2001

This License Is Not Transferable to Another Individual or Location

Permit to operate

establishment to another, or from one food operation to another. If you add a partner or incorporate, these actions are considered ownership changes and may require you to apply for a new permit to operate.

Retail food establishments are routinely inspected by regulatory agencies to assure they are complying with the food safety rules and regulations of the jurisdiction. Many regulatory agencies use a risk-based approach for setting inspection frequencies. Under this system, the risk a retail food establishment poses is calculated using such factors as:

- The sanitation history of the establishment
- Number of home meal replacements sold
- The amount of food preparation and handling conducted onsite
- Condition of building
- Critical violations observed.

Always cooperate with inspecting personnel. Greet the inspector when he or she arrives and ask to see their official identification.

Whenever possible, accompany the inspector during an inspection or audit.

A routine inspection or audit normally consists of three phases:

1. A pre-inspection conference where the inspector and manager, or person in charge, review previous inspection results and discuss information relevant to the current inspection.
2. The current inspection is conducted.
3. A post-inspection conference is conducted where the results of the current inspection are reviewed and discussed with the manager or person in charge.

Most inspections will focus on:

- Foods and supplies
- Personal hygiene and employee health
- Temperatures of food and food-holding equipment

- Cleaning and sanitizing procedures
- Equipment and utensils
- Water supply and waste disposal
- Pest control and other aspects of the operation that might compromise food safety.

If a critical violation such as temperature abuse or cross contamination is discovered during an inspection, corrective action must be taken immediately. Failure to do so could result in fines and possible closure of the establishment. Lesser violations should be corrected as soon as possible, but always prior to the next inspection or audit.

Food managers should periodically conduct self-inspections of their establishment to assure proper sanitation and food safety. During a self-inspection, the manager should look for signs of contamination, improper food-handling practices, and other conditions that may put food safety in jeopardy.

Federal Agencies

The primary federal agencies that protect our food supply and their functions are provided in the table below.

Federal Agencies	
Name	**Primary Function or Duties**
FDA **Food and Drug Administration**	• Regulates the processing, manufacturing, and interstate sale of many food items, except for meat, poultry, and egg products • Sets standards with respect to composition, quality, labeling, and safety of foods and food additives • Publishes the *FDA Food Code* • Protects the public's health by preventing the adulteration and misbranding of food • Assures food shipped in interstate commerce is safe, pure, wholesome, sanitary, and honestly packaged and labeled • Maintains a list of interstate certified shellfish shippers and a list of interstate milk shippers.

(Continued)

Federal Agencies (continued)

Name	Primary Function or Duties
USDA **U.S. Department of Agriculture**	• Inspects domestic and imported meats, poultry, and processed meat and poultry products. USDA maintains a list of approved facilities for meat and poultry processing. • Conducts voluntary grading services for red meats, poultry and eggs, dairy products, and fruits and vegetables.
USDC **U.S. Department of Commerce**	• Develops grade standards for processed fishery products.
NMFS **National Marine Fisheries Service**	• Division of the USDC • Provides voluntary inspection service for processed fishery products • Maintains a list of approved fish processors and fishery products on a permanent basis.
EPA **Environmental Protection Agency**	• Created to prevent, control, and reduce air, land, and water pollution • Regulates the use of toxic substances (pesticides, sanitizers, and other chemicals), monitors compliance, and provides technical assistance to states.
CDC **Centers for Disease Control and Prevention**	• Responsible for protecting public health through the prevention and control of diseases • Supports foodborne disease investigations and prepares annual summaries and statistics on outbreaks of foodborne diseases transmitted through food and water.

(Continued)

Federal Agencies (Continued)	
Name	**Primary Function or Duties**
OSHA **Occupational Safety and Health Administration**	• Created in the U.S. Department of Labor to enforce the Occupational Safety and Health Act. It enforces health standards and regulations on safety, noise, and other workplace-related hazards.
FTC **Federal Trade Commission**	• Involved in enforcing various laws regarding marketing practices and national advertising of foods and other products.

The FDA publishes the *FDA Food Code*, which serves as a model for retail food programs regulated by federal, state, local, and tribal agencies. The *FDA Food Code* is not a federal law or federal regulation, and it does not replace existing retail food laws. The code is more like a set of recommendations put forth to promote food safety and sanitation nationwide.

Other National Food Safety Related Organizations

The Conference for Food Protection (CFP) is a nonprofit organization that represents the major stakeholders in retail food safety. Its membership includes representatives of the food industry, government, academic community, professional organizations, and consumer groups. It meets at least every other year to identify food safety problems, make recommendations, and implement practices to assure food safety. The CFP has no formal regulatory authority, yet over the years it has been able to effectively influence the content of model retail food laws and regulations.

The mission of the CFP is to promote food safety and consumer protection.

Other organizations also help protect our food supply. Many of the following organizations have good information and training materials:

- The American Public Health Association (APHA)
- The Association of Food and Drug Officials (AFDO)
- The Food Marketing Institute (FMI)
- The Frozen Food Industry Coordinating Committee (FFICC)
- The Food Research Institute (FRI)
- Institute of Food Technologists (IFT)
- National Conference on Interstate Milk Shipments (NCIMS)
- International Association for Food Protection (IAFP)
- The National Environmental Health Association (NEHA)
- The National Pest Management Association (NPMA)
- The National Restaurant Association (NRA)
- NSF International (NSF)
- The National Shellfish Sanitation Program (NSSP)
- Underwriters Laboratories, Inc. (UL)
- World Health Organization (WHO).

Inspection for Wholesomeness, Grading and Generally Recognized as Safe

Inspection for wholesomeness refers to an examination of domestic and imported meat, poultry, and egg products to assure they are wholesome and free from adulteration. These inspections are required by law and are conducted by veterinarians and technicians who work for the Food Safety and Inspection Service (FSIS) of the USDA. The USDA can also be involved with microbial testing of foods. *E. coli O157:H7* has been declared an adulterant in raw ground beef, and no amount of these bacteria are permitted in ground beef by law. A national microbial sampling plan has been implemented to determine the prevalence of *E. coli O157:H7* bacteria in ground beef. Under the sampling plan, USDA officials take samples of ground beef from federally inspected plants and retail food establishments and test them for *E. coli O157:H7* bacteria. If a sample is confirmed positive for *E. coli 0157:H7,* the retail food establishment may be asked to recall the product and communicate this information to the public. It is important to develop a written procedure (i.e., SOP) for ground beef sampling and to assure measures are in place to identify different lots of ground beef.

Grading refers to the process of evaluating foods relative to specific, defined standards in order to assess its quality. Grading is a voluntary activity. However, most food processors participate in grading programs because their customers prefer to buy products that have met specific quality standards. More information on grading is provided in Chapter 4 of this book

Substances used in foods for years and with no apparent ill effects are commonly identified as **Generally Recognized As Safe (GRAS)**. Examples of substances that fall into the GRAS category are spices, natural seasonings, flavoring materials, fruit and beverage acids, baking powder chemicals, and drying agents. The purpose of establishing a GRAS list is to recognize the safety of basic substances without the requirement for rigorous safety testing.

Food Recall

The FDA can request a food recall if the agency has determined a potential hazard exists. Food recalls are voluntary with the possible exception of a recall involving infant formula. In the vast majority of cases, food processors voluntarily remove products from the marketplace to keep consumers safe. However, if a firm does not agree with an FDA request for a voluntary recall, the agency may issue public press releases and proceed to seize the product in order to remove it from commerce.

There are three types of food recalls. Class I, Class II, or Class III can be issued depending on the severity of the health risk. Retail food establishments should have a policy in place to handle food recalls. Recalled foods should be promptly removed from retail shelves and cannot be sold.

Types of Food Recalls

- Class I — Foods that may cause serious adverse health consequences
- Class II — Foods that would result in a temporary or reversible health problem
- Class III — Foods that are not likely to cause danger to health.

Food Labeling

Nutrition and Ingredient Labeling

The USDA and the FDA share responsibility for enforcing the Nutrition Labeling and Education Act. This law provides information on specific nutritional guidelines and defines what needs to be on a nutrition label. This label, which provides information for processed foods on protein, fat, carbohydrate, and mineral content helps customers make informed choices when selecting foods.

Although not mandatory at this time, the USDA has published guidelines for demonstrating voluntary compliance to nutritional labeling requirements. Supermarkets are encouraged to provide consumers accurate nutritional information using point-of-purchase product labels, brochures or signage.

Retail establishments that prepare and package product in the store also must provide a label for the consumer. All foods packaged in a retail food establishment must be labeled as specified in the law (21 CFR 101 - Food Labeling, and 9 CFR 317 Labeling, Marking Devices, and Containers). The label must include information such as the name of the food, a list of ingredients, and the quantity of ingredients.

Meltaway/French Mint Combo 51/2 oz.

Nutrition Facts

Serving Size: 42.0g, 5 pieces
Serving Per Container: 4

Amount Per Serving

Calories 248 Calories from Fat 150

	% Daily Value*
Total Fat 17g	**26%**
Saturated Fat 10g	**50%**
Cholesterol Less then 5mg	**0%**
Sodium 50mg	**2%**
Total Carbohydrate 22g	**7%**
Dietary Fibers 1g	**4%**
Sugars 20g	
Protein 2g	

Vitamin A 0%	Vitamin C 0%
Calcium 6%	Iron 2%

* Percent Daily Values are based on a 2,000 calorie diet. Your daily values may be higher or lower depending on your calorie needs:

	Calories	2,000	2,500
Total Fat	Less than	65g	60g
Sat Fat	Less than	20g	25g
Cholesterol	Less than	300mg	300mg
Sodium	Less than	2,400mg	2,400mg
Total Carbohydrate		300g	375g
Dietary Fiber		25g	30g

19361

The nutrition label
(Source: www.nal.usda.gov/fnic/)

Labeling is critical when dealing with common allergens.

Safe Food-handling Label

A safe food-handling instruction label provides helpful food-handling information for the consumer on all raw or partially precooked (not ready-to-eat) meat and poultry. The label was designed to educate the consumer for storage, preparation, and handling of raw meat and poultry products in the home.

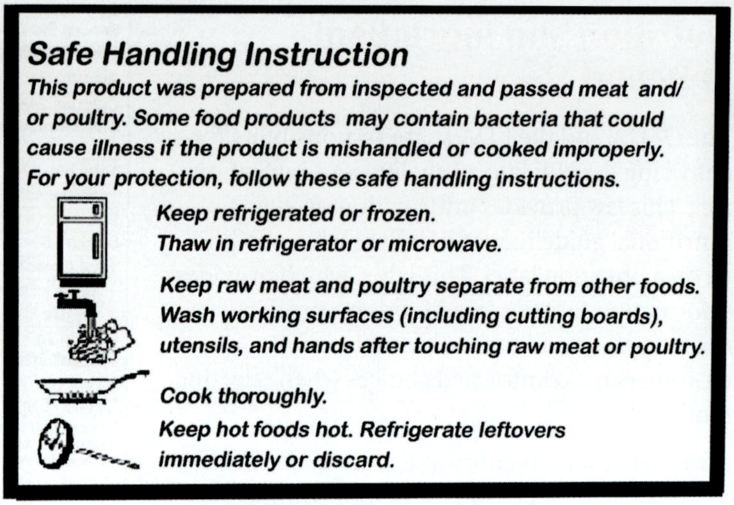

USDA safe handling instructions label
(Source: http://www.usda.gov/agency/oc/design/art_symbols.html)

Product Dating Labels

A product dating label is a date stamped on a food package to help a retail food establishment know how long to display the product for sale. It can also help customers understand the time limit to purchase or use the product at its best quality. Date labels are on almost all food products. Date labels are usually found on perishable, potentially hazardous foods such as meat, poultry, eggs, and dairy items.

Product labels with date

There are many types of dating labels that can be used. A sell by date tells the retail food establishment how long to display the product for sale. Food products should be removed from the retail shelves immediately after the sell by date, and customers should buy the product before the date expires. A best if used by (or before) date is recommended for the best flavor or quality. If handled properly, the product quality will last two to five days past the sell by date. A use by date is the last date recommended for the use of the product while at peak quality. The manufacturer of the product usually determines the date by using quality control and shelf-life tests. Federal laws regulate date marking of infant formula products.

Recent Initiatives in Food Safety

National Food Safety Initiative

From farm to table, the federal government's National Food Safety Initiative is working to reduce the incidence of foodborne illness by strengthening and improving food safety practices and policies. The initiative seeks to:

- Expand education efforts aimed at improving food handling in homes and retail outlets.
- Improve inspections and expand monitoring efforts.
- Increase research to develop new and more rapid methods to detect foodborne pathogens and to develop preventive techniques.
- Improve intergovernmental communications and coordination of responses to foodborne outbreaks. FDA, CDC, USDA, and EPA will form a new intergovernmental group to improve federal, state, and local responses to outbreaks of foodborne illnesses.
- Determine the frequency and severity of foodborne disease.
- Determine the proportion of common foodborne diseases that result from eating specific foods.
- Describe the epidemiology of new and emerging bacterial, parasitic, and viral foodborne pathogens.

The Fight BAC!™ Campaign

The Fight Bac!™ Campaign is an educational program developed by industry, government, and consumer groups to teach customers about safe food handling. The focal point of the campaign is BAC (short for bacteria) a character used to make customers aware of the invisible germs that cause foodborne illness. Even though customers cannot see, smell, or taste BAC, it and millions more germs like it are in and on food and food-contact surfaces. Fight Bac!™ Campaign uses media and community outreach programs to teach

Fight Bac!™ Logo

customers how to keep food safe from harmful bacteria through four simple steps: wash hands and surfaces often, prevent cross contamination, cook foods to proper temperature, and refrigerate foods quickly. The partners in the Fight Bac!™ Campaign use newspaper articles, public service announcements, printed brochures, and school curricula to teach customers how to lower their risk of foodborne illness.

Consumer Advisory

The food establishment can satisfy the consumer advisory requirements by providing both a disclosure and a reminder.

Food establishments that sell or serve raw or undercooked animal foods or ingredients for human consumption must inform their customers about the increased risk associated with eating these kinds of foods. The disclosure is used to make customers aware of the dangers of raw or undercooked foods and of items that either contain or may contain raw or undercooked ingredients. The reminder is a notice about the relationship between thorough cooking and food safety.

Disclosure is satisfied when items are described such as oysters on the half-shell (raw oysters), and raw-egg Caesar salad; or items are asterisked to a footnote that states the items:

- Are served raw, or undercooked, or
- Contain (or may contain) raw or undercooked ingredients.

The reminder is satisfied when the items requiring disclosure are asterisked to a footnote that states information about:

- The safety of these items, written information is available upon request
- Eating raw or undercooked meats, poultry, seafood, shellfish, or eggs, which may increase your risk of foodborne illness
- Increased risk of foodborne illness when the above products are eaten if you have certain medical conditions.

Disclosure of raw or undercooked animal-derived foods or ingredients and reminders about the risk of consuming such foods should appear at the point where the customer selects the food. Both the disclosure and the reminder need to accompany the information from which the customer makes a selection. That information could appear in many forms such as display signage, package labels, and consumer brochures.

Consumer advisories may be tailored to be product specific if a food establishment either has a limited menu or offers only certain animal-derived foods in a raw or undercooked, ready-to-eat form. For example, a raw bar serving molluscan shellfish on the half shell, but no other raw or undercooked animal food, could elect to confine its consumer advisory to shellfish.

Check It Out

Back to the Story... Food allergens have become a very important concern in recent years. Allergic reactions to peanuts can be especially life threatening. In this case, the candies were not properly labeled to warn the consumer peanuts were part of the ingredients. In turn, the manufacturer initiated a food recall. But, the response and responsibility of the retail food manager becomes very important. It will be their job to remove products from the retail shelves as quickly as possible and communicate the risks of the product that has already been purchased. Our federal government requires manufactured goods be identified with a product code that helps facilitate the identity of the recalled product. Once the recall begins, check your shelves and storage facilities and be sure to remove the product from sale.

- ✓ Federal, state, and local food safety regulations are important for maintaining a safe food supply. State and local health departments are critical when it comes to monitoring and enforcing each state or local food code. They evaluate food establishments to help assure delivery of safe food to customers.
- ✓ Food establishments and regulatory agencies should work together to protect the health of customers and meet their expectations for food quality and safety.

Quiz 11 (Multiple Choice)

Please choose the BEST answer to the questions.

1. The federal agency responsible for safe working conditions such as noise and exposure to hazardous chemicals is the:
   a. Environmental Protection Agency (EPA).
   b. Food and Drug Administration (FDA).
   c. Occupational Safety and Health Administration (OSHA).
   d. Centers for Disease Control and Prevention (CDC).

2. Which of the following regulatory agencies will have the most influence on food-handling practices for foods prepared in a retail food establishment?
   a. State and local health department or department of agriculture.
   b. U.S. Food and Drug Administration (FDA).
   c. U.S. Department of Agriculture (USDA).
   d. Centers for Disease Control and Prevention (CDC).
3. When an inspector from your local regulatory agency finds a violation of code rules, the manager and supervisor should:
   a. Decide on a time to meet and discuss the problem.
   b. Insist this has never caused a problem before.
   c. Wait for a report to verify the problem.
   d. Immediately begin to fix the problem.
4. If a customer reports a suspected foodborne illness thought to be from a product in your area, you should immediately notify the manager and:
   a. Wait for a decision on action to take.
   b. Remove suspected product from use or sale.
   c. Wait until you have two or more related complaints.
   d. Notify the local health department.
5. Which federal agency prepares the model *Food Code*?
   a. Food and Drug Administration (FDA).
   b. U.S. Department of Agriculture (USDA).
   c. Environmental Protection Agency (EPA).
   d. Centers for Disease Control and Prevention (CDC).

References/Suggested Reading

Food Safety and Inspection Service, U.S. Department of Agriculture, Washington, DC. 20250-3700, August 2000, http://www.fsis.gov/oa/pubs/dating.htm

IFT Publication (1992). *Government Regulation of Food Safety: Interaction of Scientific and Societal Forces.* Food Technology. 46(1).

Labuza, T. P., and Baisier (1992). *The Role of the Federal Government in Food Safety.* Critical Reviews in Science and Nutrition. 31(3). 167-176.

Potter, N.N., and J. H. Hotchkiss (1995). *Food Science.* Chapman and Hall. New York, NY.

Glossary

Abrasive cleaners - cleaning compounds containing finely ground minerals used to scour articles; can scratch surfaces.

Acceptable level - within an established range of safety.

Accessible - easy to reach or enter.

Acid - a substance with a pH of less than 7.0.

Additives - natural and man-made substances added to a food for an intended purpose (such as preservatives and colors) or added unintentionally (such as pesticides and lubricants).

Adulterated - the deliberate addition of inferior or cheaper material to a supposedly pure food product in order to stretch out supplies and increase profits.

Aerobe - an organism, especially a bacterium, that requires oxygen to live.

Air curtain - a window or doorway equipped with jets that force air out and across the opening to keep flying insects from entering.

Air-dry - to dry in room air after cleaning, washing, or sanitizing.

Air gap - an unobstructed, open vertical distance through air that separates an outlet of the potable water supply from a potentially contaminated source like a drain.

Alkaline - a substance that has a pH of more than 7.0.

Ambient - a recommended air temperature used to transport or to store perishable foods.

Anaerobe - an organism, especially a bacterium, that does not require oxygen or free oxygen to live.

Anisakiasis - disease caused by the *anisakis* parasite.

Anisakis - roundworm parasite found in fish.

Aseptic packaging - a method in which food is sterilized or commercially sterilized outside the can and then aseptically placed in previously sterilized containers which are then sealed in an aseptic environment. This method may be used for liquid foods such as concentrated milk and soups.

At risk - the term used to describe individuals such as infants, children, pregnant women, and those with weakened immune systems, for whom foodborne illnesses can be very severe, even life threatening.

Backflow - flow of contaminated water into potable supply; caused by back-pressure.

Backroom - a storage area for excess products kept on hand to restock the sales floor as needed.

Back-siphonage - a form of backflow that can occur when pressure in the potable water supply drops below pressure in the flow of contaminated water.

Bacteria - single-celled microscopic organisms.

Bactericide - substance that kills bacteria.

Bacterium - one microorganism.

Bagger - a retail clerk or associate who bags customers' purchases in plastic or paper bags to suit customers.

Bagging - a process of properly, carefully packing a customers' purchases in plastic or paper bags to suit customers.

Bake-off - an in-store baking process using frozen doughs and products to prepare fresh products, i.e., fresh rolls, bread, doughnuts or other pastries.

Bale - a large bundle of cardboard that is recycled.

Baler - a device used to compact and bind corrugated cardboard into bales for recycling.

Bar code - a unique identification code on products, pallets and coupons. The code is read by an electronic scanner for receiving, ordering and inventory control purposes.

Bi-metallic stem thermometer - food thermometer used to measure product temperatures.

Binary fission - the process by which bacteria grow. One cell divides to form two new cells.

Biofilm - a layer of stubborn soil that develops on surfaces that are improperly cleaned or sanitized; bacteria can accumulate and grow on this surface.

Blast chiller - special refrigerated unit that quickly freezes food items.

Bloating - a damaged, swollen processed food can or glass container, which may indicate contamination, a safety hazard.

Bloom - an indication of freshness and quality as beef turns bright red when exposed to oxygen.

Bottle returns - beverage bottles returned to a retailer for recycling.

Botulism - type of food intoxication caused by *C. botulinum.*

Box cutter - a knife-like device with a razor blade used to open boxes.

Braise - to cook meat by browning it in fat, then simmering it in a covered pan with a little liquid.

Breaking down - removing products from a case in order to clean and sanitize it; also, removing component parts of a piece of equipment, such as a slicer or a grinder to clean and sanitize it.

Bulk produce - loose, unpackaged, fresh produce that customers select themselves.

Bulk product - unpackaged, fresh products displayed in bins in large quantities and sold by the piece or the pound, such as grains, candy and snacks.

Calibrate - to determine and verify the scale of a measuring instrument with a standard. Thermometers used in food establishments are commonly calibrated using an ice slush method (32°F or 0°C) or a boiling point method (212°F or 100°C).

Campylobacter jejuni - a microaerophilic non-sporeforming bacterium that causes a foodborne illness.

CAP - controlled atmosphere packaging.

Cardiopulmonary resuscitation (CPR) - first-aid technique to apply to persons who have stopped breathing.

Case, refrigerated - a refrigerated display unit for perishable products, such as dairy products or ice cream.

CBT - computer-based training

Centralized prepackaging - a centralized facility that processes prepackaged perishables and ships the goods to stores.

Channel of distribution - the producers and distributors of products from a farm to the table, a path that includes a grower, producer, manufacturer, broker, wholesaler, store and consumer.

Checker - a front-end employee who rings up, totals and collects for a customer's order. Also known as a cashier.

Checking - the process of recording customer purchases, taking payment, making change, processing coupons, bagging and all other functions inherent to the front end operation.

Checkout or checkstand - a fixture in a supermarket where customer transactions occur and where the register, scanner, and bags are located. Also known as the front end.

Chemical sanitizers - products used on equipment and utensils after washing and rinsing to reduce the number of disease-causing microbes to safe levels.

Ciguatoxin - a toxin from reef-feeding fish that causes foodborne illness.

Clean - free of visible soil but not necessarily sanitized; surface must be cleaned before it can be sanitized.

Cleaning - removal of visible soil but not necessarily sanitized. A surface must be cleaned before it can be sanitized.

Cleaning agent - a chemical compound formulated to remove soil and dirt.

Clean-in-Place (CIP) - equipment designed to be cleaned without moving, usually large or very heavy.

Code - a systematic collection of regulations, or statutes, and procedures designed to protect the public.

Code date or dating - a date and source code printed on an item to indicate its shelf life. Date codes simplify rotation and help prevent the sale of off-quality or spoiled products. See pull date.

Coffin case or coffin freezer - a waist-high fixture used to display frozen food, with a transparent door or no door for easy access. See upright freezer.

Cold-holding - refers to the safe temperature range (less than 41°F or 5°C) for maintaining foods cold prior to service for consumption.

Cold storage - a facility that stores frozen foods and perishable items that need refrigeration or special handling.

Commingle - to combine shellstock harvested on different days or from different growing areas as identified on the tag or label, or to combine shucked shellfish from containers with different container codes or different shucking dates.

Comminuted - to reduce the size of a food mass by methods including chopping, flaking, grinding, or mincing.

Commissary - A centralized food preparation facility which distributes prepared product to stores. Also a U.S. government-subsidized, non-profit food store, operating on a military base. Operates on a non-profit basis and sells products at cost plus a small markup.

Compactor, trash - a device used to crush dry or wet garbage. Often found in many stores in two separate units–one compactor for paper and cardboard and one for all other materials.

Competency based training - a job is analyzed and broken down into steps which must be learned to gain mastery of the task; training should focus on those steps.

Computer-based training (CBT) - a self-paced, interactive, computerized tutorial or training process (see CBT).

Consumer - an end user of any product or service. A shopper, patron or customer. The final link in the chain of product distribution: manufacturing, selling, wholesaling, retailing, consuming.

Contamination - the unintended presence of harmful substances or conditions in food that can cause illness or injury to people who eat the infected food.

Controlled Atmosphere Packaging (CAP) - a product in a low-oxygen, nitrogen-rich wrap that preserves freshness.

Convenience store - a small, easy-access food store with a limited assortment. Many convenience stores also sell fast food and gasoline.

Conventional supermarket - a large, self-service, retail food store with moderate pricing and selection, and annual sales in the $2 to $8 million range. Usually includes a meat, produce, dairy and grocery department.

Cook-chill - a cooking and food preservation process for food service products. Food is cooked, packed, sealed and quick chilled in a plastic pouch and stored at temperatures below 41°F (5°C).

Cooking - the act of providing sufficient heat and time to a given food to effect a change in food texture, aroma, and appearance. More importantly, cooking ensures the destruction of foodborne pathogens inherent to that food.

Cooler - a refrigerated holding unit in a warehouse or store for perishables.

Cooling - the act of reducing the temperature of properly cooked food to 41°F (5°C) or below.

Corrosion resistant - materials which maintain their original surface characteristics under continuous use in food service with normal use of cleaning compounds and sanitizing solutions.

Coving - a curved sealed edge between the floor and wall that makes cleaning easier and inhibits insect harborage.

Critical Control Point (CCP) - means a point or procedure in a specific food system where loss of control may result in an unacceptable health risk.

Critical limit - the maximum or minimum value to which a physical, biological, or chemical parameter must be controlled at a critical control point to minimize the risk that the identified food safety hazard may occur.

Cross-connection - any physical link through which contaminants from drains, sewers, or waste pipes can enter a potable water supply.

Cross contamination - transfer of harmful organisms between items by direct or indirect contact.

Cryovac - a proprietary term for vacuum packaging material, which has entered the language to mean all vacuum packaging, like "Kleenex" or "Band-aid."

Cumulative - increasing in effect by successive additions. For example, hot food must be cooled from 135°F (57°C) to 70°F (21°C) within two hours; it must reach 41°F (5°C) within an additional four hours to prevent bacterial growth. Therefore, the cumulative time the food is in this danger zone equals a total of six hours or less.

Cutting - opening or sampling a product to evaluate its appearance, flavor, quality and/or consistency.

Dairy/deli case extender - an insulated container display attached to a refrigerated case that extends into an aisle to stimulate impulse buys.

Danger zone - temperatures between 41°F (5°C) and 135°F (57°C).

Date code or coding - a "sell by" date stamped on a product to ensure freshness. The date assists with quality control (first-in, first-out) and proper rotation. It may also apply to affixing a "sell by" or "pull by" date on merchandise which is on display (as in the Bakery Department).

Disinfectant - destroys harmful bacteria.

Dead man's switch - activates equipment when depressed and stops if pressure is relieved.

Deli-bake - a combination in-store bakery and deli department where equipment, floor space and labor are shared, usually under common supervision.

Delicatessen - an in-store department with cooked foods, salads, cold cuts and cheeses, etc.

Demo or demonstration - a product promotion in a store with samples to eat and cooking-tip handouts or coupons.

Detergents - cleaning agent which contains surfactants used with water to break down soil to make it easier to remove.

Deviation - to diverge; to go in different directions.

Digital thermometer - a battery-powered thermometer that reveals temperature in a digital numerical display.

Disclosure - a written statement that clearly identifies the animal-derived foods which are, or can be ordered raw, undercooked, or without otherwise being processed to eliminate pathogens in their entirety, or items that contain an ingredient that is raw, undercooked, or without otherwise being processed to eliminate pathogens.

Dock - an area to receive, load and unload shipments.

Dressed fish - a whole scaled, cleaned fish, sold with or without the head.

Dry grocery - non-perishable grocery products.

Dry grocery non-foods - products that are not food, such as paper products, detergents, or pet items.

Easy-to-clean - materials and design which facilitate cleaning.

Easy-to-move - on wheels, raised on legs, or otherwise designed to facilitate cleaning.

Employee - person working in or for a food establishment who engages in food preparation, service, or other assigned activity.

Equipment - the appliances: stoves, ovens, etc.; and storage containers, such as refrigerated units used in food establishments.

Ethnic foods - products that a particular ethnic (racial, national, religious, or cultural) group favor such as Mexican, Chinese, or Kosher.

Evaluation procedures - systematically checking progress to determine if goals have been met.

Exotic produce - fruits and vegetables not grown in North America and considered exotic.

Expiration date - a manufacturer's "sell by" date stamped on products to indicate shelf life.

Facultative anaerobe - an organism that can grow with or without free oxygen.

FATTOM - an acronym used to indicate the six conditions bacteria need for growth. These conditions are Food, Acid, Temperature, Time, Oxygen, and Moisture.

FIFO - acronym for first-in, first-out, used to describe stock rotation procedures of using older products first.

FMI - Food Marketing Institute - an international trade association of independent grocers, chain stores, and wholesalers.

Food Allergen - a substance in food that causes the human immune system to produce chemicals and histamines in order to protect the body. These chemicals produce allergic symptoms that affect the respiratory system, gastrointestinal tract, skin, or cardiovascular system.

Foodborne disease outbreak - an incident in which two or more people experience a similar illness after ingesting a common food, and in which epidemiological analysis identifies the food as the source of the illness.

Foodborne illness - an illness caused by the consumption of a contaminated food.

Food - contact surface - any surface of equipment or any utensils which food normally touches.

Food establishment - an operation that stores, prepares, packages, serves, vends, or otherwise provides food for human consumption such as a restaurant, food market, institutional feeding location, or vending location.

Footcandle - unit of lighting equal to the illumination one foot from a uniform light source.

Forklift - a vehicle with projecting prongs that slide under a pallet to move merchandise in a warehouse or store.

Fresh - just picked, gathered, produced, live or unprocessed, not stale, food. A term associated with perimeter departments, including produce, deli, bakery, or floral; also, unfrozen.

Fungi - a group of microorganisms that includes mold and yeasts.

Garbage - wet waste matter, usually food product, that cannot be recycled.

Garnish - a decoration on salads, such as sprigs of watercress, lettuce or other colorful items.

Gastroenteritis - an inflammation of the linings of stomach and intestines which can cause nausea, vomiting, abdominal cramping, and diarrhea.

General merchandise (GM) - products other than food sold in supermarkets and require special buying, warehousing and servicing. GM classes are: hardiness, softness, reading/writing lines, health and beauty care (HBC) and services.

Generally recognized as safe (GRAS) - a food safety FDA term that indicates all ingredients are approved for human consumption. (See GRAS Substances.)

Germicide - a substance which kills harmful microbes and germs.

Germs - general term for microorganisms, including bacteria and viruses.

Grade - a food industry classification system or standard that indicates a quality level, such as Grade A, Prime, or Extra Fancy.

Grade standards - primarily standards of quality to help producers, wholesalers, retailers, and patrons in marketing and purchasing food products. The grade standards are not aimed at protecting the health of the consumer but rather at ensuring value received according to uniform quality standards.

GRAS substances - GRAS stands for generally recognized as safe. These are substances added to foods that have been shown to be safe based on a long history of common usage in food.

Grocery store - a retail store that sells a variety of food products, including some perishable items and general merchandise.

Grocery wholesaler - a middleman who buys food and supplies from manufacturers to resell them in smaller quantities to retailers; cooperatives and voluntaries are the two major types.

HACCP - Hazard Analysis Critical Control Point.

Hand washing - the proper cleaning of hands with soap and warm water to remove dirt, filth, and disease germs.

Harborage - shelter for pests.

Hazard - a biological, chemical, or physical agent that may cause an unacceptable consumer health risk.

Hazard analysis - to identify hazards (problems) that might be introduced into food by unsafe practices or the intended use of the product.

Hazard Analysis Critical Control Point (HACCP) - a federal guideline to ensure safe food handling and preparation from receiving to point of sale. (See HACCP.)

Heat-and-eat - a precooked food that requires heating before consumption.

Heimlich maneuver - method used to expel foreign body caught in someone's throat.

Hermetic packaging - a container completely sealed by heat against the entry of bacteria, molds, yeasts, and filth as long as it remains intact.

Highly susceptible population - persons who are more likely than other people in the general population to experience foodborne disease because they are immunocompromised, preschool age children, or older adults.

Home meal replacement - foods prepared in a store and consumed at home or in-store which require little or no preparation on the part of the consumer.

Hot-holding - refers to the safe temperature range of 135°F (57°C) and above to maintain properly cooked foods hot until served.

Housekeeping - operational procedures to ensure cleanliness, safety, sanitation and maintenance for a store or warehouse.

Immunosuppressed - an individual who is susceptible to illness because his or her immune system is not working properly or is weak.

Impermeable - does not permit passage, especially of fluids.

Individually quick frozen (IQF) - a food processing technique that freezes products in the final stage of processing. It is then wrapped and packaged for shipment.

Infection - illness caused by eating food that contains living disease-causing microorganisms.

Infestation - presence of a large number of pests.

Ingestion - the process of eating and digesting food.

Inherent - being an essential part of something.

In-store bakery - a bakery department producing baked goods in the store using one of two baking methods: scratch using ready-made mixes and frozen dough, making products according to a formula or bake-off.

In-store demonstrators - an employee or temporary person conducting tastings, samplings, and product preparations to encourage impulse buys from customers.

Insulated containers - containers used to hold hot or cold products to ensure that they will be in good condition when they get to the customer's home.

Integrated Pest Management (IPM) - a system of preventive and control measures used to control or eliminate pest infestations in food establishments.

Intoxication - illness caused by eating food that contains a harmful chemical or toxin.

Irradiation - a food preservation process that utilizes radiation to control bacteria growth and increase shelf life.

Kick plate (base) - a metal sheet, usually at the bottom of doors, for protection purposes.

Kitchenware - utensils used to prepare food.

Larva - immature stage of development of insects and parasites.

Leftovers - any food that is prepared for a particular meal and is held over for service at a future meal.

Low temp - A refrigerator that holds products at a below-freezing temperature 32°F (0°C)) or less.

Manager - the individual present at a food service establishment who supervises employees who are responsible for the storage, preparation, display, and service of food to the public.

MAP - modified atmosphere packaging.

Measuring device - usually a thermometer that registers the temperature of products and water used for sanitizing.

Media - newspapers, magazines, radio, and television.

Mesophile - microorganisms that grow best at moderate temperatures.

Microbe - a microorganism that can cause disease. Bacteria, molds and yeast that can grow on various food and equipment surfaces; the main cause of discoloration in meat and food poisoning.

Microorganism - bacteria, viruses, molds, and other tiny organisms that are too small to be seen with the naked eye. The organisms are also referred to as microbes because they cannot be seen without the aid of a microscope.

Misbranding - falsely or misleadingly packaged or labeled food; may contain ingredients not included on label or does not meet national standards for that food.

Modified Atmosphere Packaging (MAP) - a food processing technique where foods are placed in a flexible container and the air is removed from the package. Gases may be added to help preserve the food. (See MAP.)

Mold - any of various fungi that spoils food and has a fuzzy appearance.

Monitoring procedures - a defined method of checking foods during receiving, storage, preparation, holding, and serving processes.

Non food-contact surface - any area not designed to touch food.

Nutrition labeling - an accurate list of ingredients printed on food, beverage and drug labels.

Occupational Safety and Health Administration (OSHA) - a federal agency that sets work place safety standards and inspects facilities for safe working conditions.

OJT - on-the-job training.

Onset time - the period between eating a contaminated food and developing symptoms of a foodborne illness.

On-the-job-training - an instructional method to teach employees skills as they work. (See OJT.)

Open dating - a date stamped or printed on the label of perishable items to indicate a pull date (a date by which the item must be sold or removed from the shelf) or pack date (the date the item was packaged). Clear, readable dates that are printed on labels.

Organically grown - an imprecise term that means a grower did not use chemicals or that a processor did not use preservatives in the product.

Ovenable - a food ready to be heated, either in an oven or microwave.

Overstock - an excessive amount of product purchased in anticipation of increased sales volume.

Over-wrap - to wrap a plastic container in cellophane to prevent tampering.

Pack date - the date on which a product was made or packaged for sale.

Palatable - food that has an acceptable taste and flavor.

Pallet - a standard-sized base for assembling, sorting, stacking, handling and transporting goods as a unit. The industry standard is GPC-spec-4-way entry, 48 inches x 40 inches hardwood pallets.

Parasite - an animal or plant that lives in or on another and from whose body it obtains nourishment.

Parts per million (ppm) - unit of measure for water hardness and chemical sanitizing solution concentrations.

Pasteurization - a low heat treatment used to destroy disease-causing organisms and/or extend the shelf life of a product by destroying organisms and enzymes that cause spoilage.

Pathogenic - capable of causing disease; harmful; any disease-causing agent.

Perishables - foods requiring refrigeration or special handling because they spoil easily, such as meat, seafood, produce, deli, bakery and dairy.

Personal hygiene - health habits including bathing, washing hair, wearing clean clothing, and proper hand washing.

Pest control operator - licensed individual or certified technician who provides pest control services.

pH - the symbol that describes the acidity or alkalinity of a substance, such as food.

Physical hazard - particles or fragments of items not supposed to be in foods.

Point-of-purchase - the locations within a retail store where a customer purchases products.

Point of sale (POS) - the place in a retail store where products are scanned through the register system, data is collected, and sales are tendered. POS also describes sales data generated by checkout scanners.

Potable water - water that is safe to drink.

Potentially hazardous food - a food that is natural or man-made and is in a form capable of supporting the rapid and progressive growth of infectious and toxin-producing microorganisms. The foods usually have high protein and moisture contents and low acidity.

Pounds per square inch (psi) - amount of pressure per square inch.

Premises - physical environment of the food establishment; grounds, interior, and exterior of building(s).

Preventive measures - procedures that keep foods from becoming contaminated.

Proof box - a piece of equipment, in which heat and humidity are controlled in order for dough to rise in preparation for baking.

Psychrophiles - microorganisms that grow best at cold temperatures.

Pull date - The date by which a product must be either sold or pulled from a shelf.

Quaternary ammonium - a chemical sanitizing compound that is relatively safe for skin contact and is generally noncorrosive; effective in both acid and alkaline solutions.

Reach-in case - a refrigerated display case with a self-service door used for perishable products.

Ready-to-eat foods - products that are in a form that is edible without washing, cooking, or additional preparation by the food establishment or the patron.

Receiver - an authorized associate of a warehouse or retail store who receives and checks deliveries for condition and accurate amount. The first handler of the delivery receipt or invoice.

Receiving - a door or dock of a warehouse or store designated for receiving merchandise from a supplier. The procedure for physically and legally accepting a shipment of product.

Receiving clerk - see receiver.

Receiving log - the record or listing of products received with appropriate entries.

Reclaimed goods - Unsalable product at the time of delivery that is returned to a wholesaler/vendor for reclamation.

Reconstitute - to combine dehydrated foods with water or other liquids to bring back to its original state; example, addition of water to powdered milk.

Refuse - trash, rubbish, waste.

Reheating - the act of providing sufficient heat (at least 165°F or 74°C) within a two-hour time period to ensure the destruction of any foodborne pathogens that may be present in that cooked and cooled food.

Reminder - a written statement concerning the health risk of consuming animal foods raw, undercooked, or without otherwise being processed to eliminate pathogens.

Returns - unsold, damaged, or defective merchandise sent to a supplier or distributor for credit or refund.

Rework - perishables: To crisp or trim a product that looks case-worn. Grocery: To re-affix labels. To refine a category or shelf set.

Re-wraps - products that are removed, reconditioned (if salable) and displayed with limited sell-by dates.

Risk - the chance of injury, damage, or loss.

Rotation - a shelf-stocking procedure that ensures first-in, first-out by pulling older stock forward and placing newer stock at the back during restocking.

Rotisserie - a rotating grill with an electrically turned spit that cooks meats.

Safety cutter - a case cutter used to open cases of product.

Salvage - product containers/shippers (bales, pallets, containers) that must be returned or recycled to defray operational costs.

Sanitary - healthful and hygienic. The number of harmful microorganisms and other contaminants have been reduced to safe levels.

Sanitation - maintenance of conditions which are clean and promote good health.

Sanitizer - approved substance or method to use when sanitizing.

Sanitizing - application of an agent that reduces microbes to safe levels.

Sealed - closed tightly.

Segregation - locating general merchandise products (GM) in a well-defined area of a store rather than in aisles next to or across from food products.

Selectivity - chemical sanitizers, especially quats, may kill only certain organisms and not others.

Sensitive ingredient - an ingredient that is prone to support bacterial growth.

Shelf life - Time period a product can be expected to maintain maximum quality and freshness.

Shelf stable - A processed food product that remains safe to eat without refrigeration.

Shellfish - an aquatic animal, i.e., clams, oysters, mussels, crabs, lobsters, shrimp, squid and octopus which has a shell; crustaceans or mollusks.

Shellstock - raw, in-shell molluscan shellfish.

Shiga toxin-producing *Escherichia coli* - any *E. coli* capable of producing Shiga toxins (also called verocytotoxins or "Shiga-like" toxins). This includes, but is not limited to, *E. coli* reported as serotype O157:H7, O157:NM, and O157:H-.

Shucking - a process of opening shellfish, such as oysters, clams, mussels, etc.

Silicate - salt or ester derived from silica; a hard, glassy material found in sand.

Single-use articles - items intended for one use and then discarded, such as paper cups or plastic eating utensils.

Slacking - the process used to moderate the temperature of a food such as allowing a food to gradually increase from -10°F (-23°C) to 25°F (-4°C) in preparation for deep-fat frying or to facilitate even heat penetration during the cooking or reheating of a previously block-frozen flood such as spinach.

Sneeze guard - a clear, solid barrier which partially covers food in self-service areas to keep customers from coughing, sneezing, or projecting droplets of saliva directly onto food.

Soil - dirt and filth.

Sous-vide - a European food-packaging technique where a prepared product is placed in individual pouches, cooked under a vacuum and quickly chilled. Products are frozen or refrigerated until used.

Splash contact surfaces - areas that are easy to clean and designed to catch splashed substances in work areas.

Spoilage - significant food deterioration, usually caused by bacteria and enzymes, that produces a noticeable change in the taste, odor, or appearance of the product.

Spoils - goods that cannot be sold for which a retailer receives a credit from a supplier. Also called stales.

Spore - the inactive or dormant state of some rod-shaped bacteria.

Standard of identity - Food and Drug Administration (FDA) standards for food composition.

Standard operating procedures - a comprehensive book of a company's policies and procedures. Also called SOPs.

Stationary equipment - equipment that is permanently fastened to the floor, table, or countertop.

Sulfites - preservatives used to maintain freshness and color of fresh fruits and vegetables; subject to state regulations.

Supermarket - A conventional grocery store, but not a warehouse club or mass merchant, with annual sales of $2 million or more per store.

Supplier - a generic term for wholesalers, who sell to and supply retailers directly and indirectly, i.e., manufacturer, vendor, broker, reseller.

Surfactant - chemical agent in detergent that reduces the surface tension, allowing the detergent to penetrate and soak soil loose; wetting agent.

Swells - unsalable items with expanded containers or lids signifying faulty food handling, processing or sealing.

Tableware - plates, cups, bowls, etc.

Task analysis - examining a job task to determine what it takes to do the job.

Temperature abuse - allowing foods to remain in the temperature danger zone [41°F (5°C) and 135°F (57°C)] for an unacceptable period of time.

Temperature danger zone - temperatures between 41°F (5°C) and 135°F (57°C) at which bacteria grow best.

Test kit - device that accurately measures the concentration of sanitizing solutions to ensure they are at proper levels.

Thawing - Defrosting products, also called slacking.

Thermometer - a device that measures temperatures.

Thermophiles - microorganisms that grow best at hot temperatures.

Time and temperature monitoring (TTM) devices - label-sized, chemical-filled packets that are attached to shipping cartons and indicate when the cartons have been exposed to fluctuating temperatures.

Toxin - a poisonous substance produced by microorganisms, plants, and animals and which causes various diseases.

Toxin-mediated infection - illness caused by eating a food that contains harmful microorganisms that produce a toxin once they get inside the human intestinal tract.

Trainee - an employee participating in a company-sponsored training program.

United States Department of Agriculture (USDA) - a federal agency that oversees food production and inspection. The USDA establishes grade standards for commodities, conducts agricultural research and makes results available, administers food programs (i.e., food stamps), and distributes food.

United States Department of Commerce (USDC) - a federal agency that oversees trade and competition. The USDC establishes grade standards for seafood commodities, conducts agricultural research, and makes results available.

United States grade stamp - signifies that a product is clean, safe and wholesome, and has been produced in an acceptable establishment, with the appropriate equipment, under the supervision of federal inspectors. It also indicates the product is of a specific grade.

Unsalables - products unworthy of sale, e.g., damaged, out of date, spoiled.

Upright freezer - an upright refrigerated display unit with doors used for merchandising frozen foods.

USDA Grade - United States Department of Agriculture grades that relate to a specified quality of product. Grade denotes quality and USDA denotes product inspected for wholesomeness.

Utensil - a food-contact implement or container used in the storage, preparation, transportation, dispensing, sale, or service of food, such as kitchenware or tableware that is multi-use, single-service, or single-use; gloves used in contact with food; temperature sensing probes of food temperature-measuring devices; and probe-type price or identification tags used in contact with food.

Vacuum breaker - designed for use under a continuous supply of pressure. Spring loaded device to operate after extended periods of hydrostatic pressure.

Vacuum Packaging - a packaging process in which air is removed from a package as it is sealed.

Vegetative state - the active state of a bacterium where the cell takes in nourishment, grows, and produces wastes.

Ventilation - air circulation that removes smoke, odors, moisture, and grease-laden vapors from a room and replaces them with fresh air.

Verification - to prove to be true by evidence, usually a record of time and temperatures of food from receiving to serving or vending.

Viruses - any of a group of infectious microorganisms that reproduce only in living cells. They cause diseases such as mumps and Hepatitis A and can be transmitted through food.

Warewashing - the cleaning and sanitizing of utensils and food-contact surfaces of equipment.

Water activity (A_W) - a measure of the free moisture in a food. Pure water has a water activity of 1.0, and potentially hazardous foods have a water activity of 0.85 and higher.

Waxing - applying an edible wax to some fruits and vegetables to help maintain a fresh, bright appearance and to preserve product quality.

Wetting agent - a substance which breaks down the soil to allow water and soap or detergent to liquefy and remove dirt and grease.

Whole-muscle, intact beef - whole-muscle beef that is not injected, mechanically tenderized, reconstructed, or scored and marinated, from which beef steaks may be cut.

Wholesome - something that is favorable to or promotes health.

Yeast - type of fungus that is not known to cause illness when present in foods but can cause damage to food products and will change taste; useful in making products such as bread and beer.

Appendix A

Answers to End of Chapter Questions

Chapter 1 Answers

1) a
2) d
3) c
4) d
5) d

Chapter 2 Answers

1) d
2) c
3) b
4) a
5) d

Chapter 3 Answers

1) c
2) b
3) a
4) b
5) a

Chapter 4 Answers

1) b
2) d
3) a
4) d
5) b

Chapter 5 Answers

1) c
2) d
3) a
4) b
5) d

Chapter 6 Answers

1) a
2) b
3) a
4) c
5) d

Chapter 7 Answers

1) a
2) a
3) b
4) a
5) c

Chapter 8 Answers

1) c
2) a
3) c
4) a
5) b

Chapter 9 Answers

1) c
2) d
3) b
4) b
5) a

Chapter 10 Answers

1) b
2) d
3) b
4) b
5) a

Chapter 11 Answers

1) c
2) a
3) d
4) b
5) a

Appendix B

Suggested Web Sites

The following is a list of suggested web sites to visit where you can find additional information on the topic you are studying. The list is broken into categories with chapter references. They are listed in the following manner:

(Chapter References) Name of Organization – web page URL

Accident Prevention, Emergency Response, and Worker Safety Organizations

(8) American College of Emergency Physicians – www.acep.org

(8) American Red Cross – www.redcross.org

(8) Health Answers Medical Reference Library – www.healthanswers.com

(8) Injury Control Resource Information Network – www.injurycontrol.com

(8) Medicine Net Company – www.medicinenet.com

(8) National Safety Council – www.nsc.org

(8,11) Occupational Safety and Health Administration (OSHA) – www.osha.gov

(8) Safety Online – www.safetyonline.net

(8) Safety Smart Company – www.safetytalks.com

(8) Vermont Safety Information Resources, Inc.– www.hazard.com

Equipment Companies

Equipment Suppliers for Retail Food Establishments

(5) AK Steel Stainless Steel – www.aksteel-com/markets/stainless_steels.asp

(7) B&G Equipment – www.bgequip.com

(6) Duke Manufacturing Company – www.dukemfg.com

(5-7) Hobart Corporation – www.hobartcorp.com

(5) Hussmann Company – www.hussmann.com

(5) Katchall Industries International – www.KatchAll.com

(5) Shat-R-Shield Inc. – www.shat-r-shield.com

(5) Specialty Steel Industry of North America – www.ssina.com

Plumbing Organizations

(7) American Society of Sanitary Engineering – www.asse-plumbing.org

(7) Plumbnet-Industry Resources – www.plumbnet.com/resources.html

Third Party Certification Organizations

(5-7,10) NSF International – www.nsf.org

(5-7) Underwriters Laboratories, Inc. – www.ul.com

Government Agencies Providing Food Safety Programs

(1-4,9-11) Centers for Disease Control and Prevention (CDC) – www. cdc.gov

(7) EPA Office of Prevention, Pesticides and Toxic Substances – www.epa.gov/internet/oppts

(All) Gateway to Government Food Safety Information – www.foodsafety.gov

(11) National Marine Fisheries Services – www.seafood.nmfs.noaa.gov

(1-4,9-11) USDA/FDA Food and Nutrition Information Center – www.nal.usda.gov/fnic/

(4-5,9-11) USDA Food Safety and Inspection Service (FSIS) – www.fsis.usda.gov

(All) United States Department of Agriculture (SDA) – www.usda.gov

(All) United States Food and Drug Administration (FDA) – www.fda.gov

(2,4,9,11) United States Environmental Protection Agency (EPA) – www.epa.gov

(10) Canadian Food Safety Sites Involving HACCP- foodnet.fic.ca/safety/safety.html

Independent Food Safety Organizations

(1,11) Conference for Food Protection – www.foodprotect.org

(10-11) Food and Agriculture Organization – www.fao.org

(10) International Food Information Council – ificinfo.health.org/

(1-4,9-11) Partnerships for Food Safety Education – www.fightbac.org

(2) The Food Allergy Network – www.foodalergy.com

(11) World Health Organization – www.who.int

Industry Food Safety Organizations

(4) America Meat Institute – www.meatami.org

(4) The American Egg Board – www.aeb.org

(All) The Food Marketing Institute – www.fmi.org

(1) The Food Marketing Institute-Retail Food Establishment Industry Facts –www.fmi.org/facts_figs/superfact

(4) The National Chicken Council – www.eatchicken.com

(4) U.S. Poultry and Egg Association – www.poultryegg.org

Pest Control

Supply Companies

(7) Actron, Inc.– www.actroninc.com

(7) Do-It Yourself Pest Control – www.doyourownpestcontrol.com

(7) Insect-O-Cutor – www.insect-o-cutor.com

(7) Killgerm/PestWest, USA – www.Pestwest.com

(7) PCO – www.pco.ca

(7) The Orkin Company – www.orkin.com

Pest Management Organizations

(7) Association of Applied Insect Ecologists – www.aaie.com

(7) National Pest Management Association – www.pestworld.org

(7) Pest Control Industry – www.pestweb.com

(7) Pest Control Technology magazine – www.pctonline.com

(7) Virginia Polytechnic Institute and University Pesticide Programs – www.vtpp.ext.vt.eduvm.cfsan.fda.gov/~mow/intro.html

Suppliers of Equipment and Products Used for Cleaning and Sanitizing

(6) All QA Products – www.allqa.com

(6) Bowerman Associates – www.ebowerman.com

(6) Champion Industries – www.championindustries.com

(6) DiverseyLever Inc. – www.diverseylever.com

(6-7) Ecolab – www.ecolab.com

(6) Paper Thermometer Company – www.mv.com/ipusers/paperthermometer/

(6-7) Steritech – www.stertech.com

(6) The Soap and Detergent Association – www.sdahq.org

Sources of Food Safety Education Information

(3) Food Safety Day – www.foodsci.purdue.edu/publications/foodsafetyday

(9) Prentice Hall – www.prenhall.com

(9) The Food Marketing Institute – www.fmi.org

(9) Partnerships for Food Safety Education – www.fightbac.org

Appendix C

Summary of Agents That Cause Foodborne Illness

CAUSATIVE AGENT (*Sporeforming bacteria)	TYPE OF ILLNESS	SYMPTOMS ONSET	COMMON FOODS	PREVENTION
Anasakis spp.	Parasitic infection	Coughing, vomiting 1 hr to 2 weeks	Raw or undercooked seafood, especially bottom-feeding fish	Cook fish to the proper temperature throughout; freeze to meet *2001 FDA Food Code* specification
**Bacillus cereus*	Bacterial intoxication or toxin-mediated infection	1) Diarrhea, abdominal cramps (8 to 16 hrs) 2) Vomiting type, vomiting, diarrhea, abdominal cramps (30 minutes to 6 hrs)	1) Diarrhea type: meats, milk, vegetables 2) Vomiting type: rice, starchy foods; grains and cereals	Properly heat, cool, and reheat foods
Campylobacter jejuni	Bacterial infection	Watery, bloody diarrhea (2 to 5 days)	Raw chicken, raw milk, raw meat	Properly handle and cook foods; avoid cross contamination
Ciguatoxin	Fish toxin, originating from toxic algae of tropical waters	Vertigo, hot/cold flashes, diarrhea, vomiting (15 minutes to 24 hrs)	Marine finfish, including grouper, barracuda, snappers, jacks, mackerel, triggerfish, reef fish	Purchase fish from a reputable supplier; cooking WILL NOT inactivate the toxin
**Clostridium botulinum*	Bacterial intoxication	Dizziness, double vision, difficulty in breathing and swallowing, headache (12 to 36 hrs)	Improperly canned foods, vacuum packed refrigerated foods, cooked foods in anaerobic mass	Properly heat process anaerobically packed foods; DO NOT use home canned foods

CAUSATIVE AGENT (*Sporeforming bacteria)	TYPE OF ILLNESS	SYMPTOMS ONSET	COMMON FOODS	PREVENTION
**Clostridium perfringens*	Bacterial toxin-mediated infection	Intense abdominal pains and severe diarrhea (8 to 22 hrs)	Spices, gravy, improperly cooled foods (especially meats and gravy dishes)	Properly cook, cool, and reheat foods
Cryptosporidium parvum	Parasitic infection	Severe watery diarrhea within 1 week of ingestion	Contaminated water, food contaminated by infected food workers	Use potable water supply; practice good personal hygiene and handwashing
Cyclospora cayetanensis	Parasitic infection	Watery and explosive diarrhea, loss of appetite, bloating (1 week)	Water, strawberries, raspberries, and raw vegetables	Good sanitation; reputable supplier
Food Allergens	An allergic reaction usually involving the skin, mouth, digestive tract, or airways	Skin: hives, rashes, and itching Mouth: swelling and itching of the lips and tongue Digestive tract: vomiting and diarrhea Airways: difficulty breathing, wheezing	Foods that contain: milk, egg, wheat, nuts and peanuts, fish, and shellfish	Packaged and prepared foods must be properly labeled if they contain common food allergens so sensitive people can avoid them
Giardia lamblia	Parasitic infection	Diarrhea within 1 week of contact	Contaminated water	Potable water supply; good personal hygiene and handwashing
Hepatitis A virus	Viral infection	Fever, nausea, vomiting, abdominal pain, fatigue, swelling of the liver, jaundice (15 to 50 days)	Foods that are prepared with human contact, contaminated water	Wash hands and practice good personal hygiene; avoid raw seafood
Listeria monocytogenes	Bacterial infection	1) Healthy adult: flu-like symptoms 2) Highly susceptible population: septicemia, meningitis, encephalitis, birth defects (1 day to 3 weeks)	Raw milk, dairy items, raw meats, refrigerated ready-to-eat foods, processed ready-to-eat meats such as hot dogs, raw vegetables, and seafood	Properly store and cook foods; avoid cross contamination; rotate processed refrigerated foods using FIFO to ensure timely use

CAUSATIVE AGENT (*Sporeforming bacteria)	TYPE OF ILLNESS	SYMPTOMS ONSET	COMMON FOODS	PREVENTION
Mycotoxins	Intoxication	1)Acute onset: hemorrhage, fluid buildup, possible death. 2) Chronic: cancer from small doses over time	Moldy grains: corn, corn products, peanuts, pecans, walnuts, and milk	Purchase food from a reputable supplier; keep grains and nuts dry; protect products from humidity
Norwalk virus	Viral infection	Vomiting, diarrhea, abdominal pain, headache, and low grade fever; onset 24 to 48 hrs	Sewage, contaminated water, contaminated salad ingredients, raw clams, oysters	Use potable water; cook all shellfish; handle food properly; meet time temperature guidelines for PHF
Rotavirus	Viral infection	Diarrhea (especially in infants and children), vomiting, low grade fever, 1 to 3 days onset; lasts 4 to 8 days	Sewage, contaminated water, contaminated salad ingredients, raw seafood	Good personal hygiene and handwashing; proper food-handling practices
Salmonella spp.	Bacterial infection	Nausea, fever, vomiting, abdominal cramps, diarrhea (6 to 48 hrs)	Raw meats, raw poultry, eggs, milk, dairy products	Properly cook foods; avoid cross contamination
Scombrotoxin	Seafood toxin originating from histamine-producing bacteria	Dizziness, burning feeling in the mouth, facial rash or hives, peppery taste in mouth, headache, itching, teary eyes, runny nose (1 to 30 minutes)	Tuna, mahi-mahi, bluefish, sardines, mackerel, anchovies, amberjack, abalone	Purchase fish from a reputable supplier; store fish at low temperatures to prevent growth of histamine-producing bacteria; toxin IS NOT inactivated by cooking
Shellfish toxins: PSP, DSP, DAP, NSP	Intoxication	Numbness of lips, tongue, arms, legs, neck; lack of muscle coordination (10 to 60 minutes)	Contaminated mussels, clams, oysters, scallops	Purchase from a reputable supplier

CAUSATIVE AGENT (*Sporeforming bacteria)	TYPE OF ILLNESS	SYMPTOMS ONSET	COMMON FOODS	PREVENTION
Shiga toxin-producing *Escherichia coli*	Bacterial infection or toxin-mediated infection	Bloody diarrhea followed by kidney failure and hemolytic uremic syndrome (HUS) in severe cases (12 to 72 hrs)	Undercooked hamburger, raw milk, unpasteurized apple cider, and lettuce	Practice good food sanitation; handwashing; properly handle and cook foods
Shigella spp.	Bacterial infection	Bacillary dysentery, diarrhea, fever, abdominal cramps, dehydration (1 to 7 days)	Foods that are prepared with human contact: salads, raw vegetables, milk, dairy products, raw poultry, non-potable water, ready-to-eat meat	Wash hands and practice good personal hygiene; properly cook foods
Staphylococcus aureus	Bacterial intoxication	Nausea, vomiting, abdominal cramps, headaches (2 to 6 hrs)	Foods that are prepared with human contact, cooked or processed foods	Wash hands and practice good personal hygiene; cooking WILL NOT inactivate the toxin
Toxoplasma gondii	Parasitic infection	Mild cases of the disease involve swollen lymph glands, fever, headache, and muscle aches. Severe cases may result in damage to the eye or the brain. (10 to 13 days)	Raw meats, raw vegetables and fruit	Good sanitation; reputable supplier; proper cooking
Trichinella spiralis	Parasitic infection from a nematode worm	Nausea, vomiting, diarrhea, sweating, muscle soreness (2 to 28 days)	Primarily undercooked pork products and wild game meats (bear, walrus)	Cook foods to the proper temperature throughout
Vibrio spp.	Bacterial infection	Headache, fever, chills, diarrhea, vomiting, severe electrolyte loss, gastroenteritis (2 to 48 hrs)	Raw or improperly cooked fish and shellfish	Practice good sanitation, properly cook foods, avoid serving raw seafood

Appendix D

Employee Health – Disease or Medical Condition Reportable Conditions and Activities

The *2001 FDA Food Code* requires food workers and food employee applicants to report to the person in charge information about their health and activities as they relate to diseases that are transmissible through food. Specifically, an employee must report when he or she:

- Is diagnosed with:
  - *Salmonella* Typhi
  - *Shigella spp.*
  - Shiga toxin-producing *Escherichia coli*
  - Hepatitis A virus.
- Has a symptom caused by gastrointestinal illness or infection such as vomiting, diarrhea, fever, sore throat, or jaundice
- Has a lesion containing pus such as a boil or infected wound that is open and draining
- Had a past illness from:
  - *Salmonella* Typhi within the past three months
  - *Shigella spp.* within the past month
  - Shiga toxin-producing *Escherichia coli* within the past month, or
  - Hepatitis A virus.
- Meets one or more of the following high-risk conditions:
  - Is suspected of causing, or being exposed to, a confirmed disease outbreak caused by *Salmonella* Typhi, *Shigella spp.*, Shiga toxin-producing *Escherichia coli* or Hepatitis A virus where food is implicated in the outbreak
  - Lives in the same household as, and has knowledge about, a person who is diagnosed with a disease caused by *Salmonella* Typhi, *Shigella spp.*, Shiga toxin-producing *Escherichia coli* or Hepatitis A virus
  - Lives in the same household as, and has knowledge about, a person who attends or works in a setting where there is a confirmed disease outbreak caused by *Salmonella* Typhi, *Shigella spp.*, Shiga toxin-producing *Escherichia coli* or Hepatitis A virus.

Exclusions and Restrictions

The person in charge must **exclude** a food employee from a food establishment if the employee:

- Is diagnosed with *Salmonella* Typhi, *Shigella spp.*, Shiga toxin-producing *Escherichia coli,* or Hepatitis A virus
- Works in an establishment that serves a highly susceptible population and he or she:
  - ▼ Is experiencing a symptom of acute gastrointestinal illness and meets a high-risk condition as described above
  - ▼ Is not experiencing a symptom of acute gastrointestinal illness but has a stool that yields a specimen culture that is positive for *Salmonella* Typhi, *Shigella spp.*, Shiga toxin-producing *Escherichia coli,* or Hepatitis A virus
  - ▼ Had a past illness for *Salmonella* Typhi within the last three months or
  - ▼ Had a past illness from *Shigella spp.*, Shiga toxin-producing *Escherichia coli* within the last month
  - ▼ Is jaundiced.
- Is jaundiced and the onset of jaundice occurred within the last seven calendar days

The person in charge shall **restrict** a food employee from working with exposed food, clean equipment, utensils, and linens; and unwrapped single-service and single-use articles, in a food establishment if the employee is:

- Suffering from symptoms such as diarrhea, fever, vomiting, or sore throat
- Not experiencing a symptom of acute gastrointestinal illness but has a stool that yields a specimen culture that is positive for *Salmonella* Typhi, *Shigella spp.*, Shiga toxin-producing *Escherichia coli* or Hepatitis A virus
- Jaundiced and the onset of jaundice occurred more than seven calendar days before and the establishment does not serve a highly susceptible population.

Removal of Exclusions and Restrictions

The person in charge of a retail food establishment may remove an exclusion if the:

- Person in charge obtains approval from the regulatory authority
- Excluded employee provides the person in charge written medical documentation from a physician, or if allowed by law a nurse practitioner or physician assistant, that specifies:
  - ▼ That the excluded person may work as a food employee in a food establishment, including an establishment that serves a highly susceptible population, because the person is free of the infectious agent of concern

- The symptoms experienced by the worker result from a chronic noninfectious condition such as Crohn's disease, irritable bowel syndrome, or ulcerative colitis
- The excluded person's stools are free of *Salmonella* Typhi, *Shigella spp.*, or Shiga toxin-producing *Escherichia coli,* whichever is the infectious agent of concern.

The person in charge of a retail food establishment may remove a restriction if the restricted worker:

- Is free of the symptoms such as diarrhea, fever, vomiting, or sore throat and no foodborne illness occurs that may have been caused by the restricted person
- Provides written medical documentation from a physician, or if allowed by law, a nurse practitioner or physician assistant, stating that:
    - The restricted person is free of the infectious agent that is suspected of causing the person's symptoms or causing foodborne illness
    - Symptoms experienced result from a chronic noninfectious condition such as Crohn's disease, irritable bowel syndrome, or ulcerative colitis
    - The restricted person's stools are free of *Salmonella* Typhi, *Shigella spp.*, or Shiga toxin-producing *Escherichia coli,* whichever is the infectious agent of concern.

Appendix E

Conversion Table for Fahrenheit and Celsius for Common Temperatures Used in Food Establishments

°F	°C		°F	°C
212	100		100	38
200	93		85	30
194	90		75	24
190	88		70	21
180	82		55	13
171	77		45	7
165	74		41	5
160	71		38	3
155	68		36	2
150	66		33	1
145	63		32	0
140	60		30	-1
135	57		29	-2
130	54		0	-18
120	49		-4	-20
110	43		-31	-35

Appendix F

Specific Elements of Knowledge Every Retail Food Protection Manager Should Know

I. Identify foodborne illness:

A. Define terms associated with foodborne illness:

1. Foodborne illness.
2. Foodborne outbreak.
3. Foodborne infection.
4. Foodborne intoxication.
5. Disease communicated by food.
6. Foodborne pathogens.

B. Recognize the major microorganisms and toxins that can contaminate food and the problems that can be associated with the contamination:

1. Bacteria.
2. Viruses.
3. Parasites.
4. Fungi.

C. Define and recognize potentially hazardous foods.

D. Define and recognize chemical and physical contamination and illnesses that can be associated with chemical and physical contamination.

E. Define and recognize the major contributing factors for foodborne illness.

F. Recognize how microorganisms cause foodborne illness.

II. **Identify time/temperature relationship with foodborne illness.**

A. Recognize the relationship between time/temperature and microorganisms (survival, growth, and toxin production) during the following stages:

1. Receiving.
2. Storing.
3. Thawing.
4. Cooking.
5. Holding/displaying.
6. Serving.
7. Cooling.
8. Storing (post-production).
9. Reheating.
10. Transporting.

B. Describe the use of thermometers in monitoring food temperatures:

1. Types of thermometers.
2. Techniques and frequency.
3. Calibration and frequency.

III. **Describe the relationship between personal hygiene and food safety.**

A. Recognize the association of hand contact and foodborne illness:

1. Hand washing–technique and frequency.
2. Proper use of gloves, including replacement frequency.
3. Minimal hand contact with food.

B. Recognize the association of personal habits and behaviors and foodborne illness:

1. Smoking.
2. Eating and drinking.
3. Wearing clothing that may contaminate food.
4. Personal behaviors, including coughing, sneezing, etc.

C. Recognize the association of the health of a food handler to foodborne illness:

1. Free of systems of communicable disease.
2. Free of infections spread through food on contact.
3. Food protected from contact with open wounds.

D. Recognize how policies, procedures, and management can contribute to and improve food hygiene practices.

IV. Describe methods for preventing food contamination from purchasing to serving.

A. Define terms associated with contamination:

1. Contamination.
2. Adulteration.
3. Damage.
4. Approved source.
5. Sound and safe conditions.

B. Identify potential hazards prior to delivery and during delivery:

1. Approved source.
2. Sound and safe condition.

C. Identify potential hazards and methods to minimize or eliminate hazards after delivery:

1. Personal hygiene.
2. Cross contamination.
    a. Food to food.
    b. Equipment and utensils contamination.
3. Contamination.
    a. Chemical.
    b. Additives.
    c. Physical.
4. Service/display–customer contamination.
5. Storage.
6. Re-service.

V. Identify and apply correct procedures for cleaning and sanitizing equipment and utensils.

A. Define terms associated with:

1. Cleaning.
2. Sanitizing.

B. Apply principles of cleaning and sanitizing.

C. Identify materials, equipment, detergent, sanitizer.

D. Apply appropriate methods of cleaning and sanitizing:

1. Manual warewashing.
2. Mechanical warewashing.
3. Clean-in-place (CIP).

E. Identify frequency of cleaning and sanitizing.

VI. Recognize problems and potential solutions associated with facility, equipment, and layout.

A. Identify facility, design, and construction suitable for food establishments:

1. Refrigeration.
2. Heating and cooling.
3. Floors, walls, and ceilings.
4. Pest control.
5. Lighting.
6. Plumbing.
7. Ventilation.
8. Water supply.
9. Wastewater disposal.
10. Waste disposal.

B. Recognize problems and potential solutions associated with temperature control, preventing cross contamination, housekeeping, and maintenance.

1. Develop and implement self-inspection program.
2. Develop and implement pest control program.
3. Develop and implement cleaning schedules and procedures.
4. Develop and implement equipment and facility maintenance program.

(Developed by the Conference for Food Protection)

INDEX